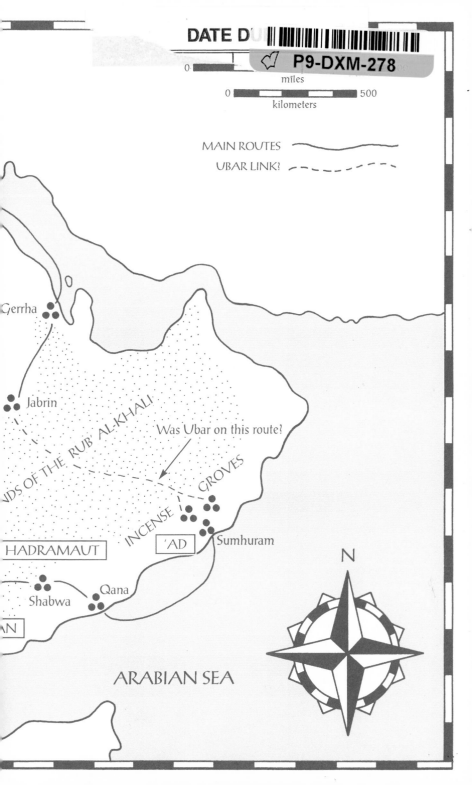

The Road to Ubar

THE ROAD
TO UBAR

Finding the
Atlantis of the Sands

Nicholas Clapp

Houghton Mifflin Company

Boston   New York

Copyright © 1998 by Nicholas Clapp

For information about permission to reproduce selections from
this book, write to Permissions, Houghton Mifflin Company,
215 Park Avenue South, New York, New York 10003.

LIBRARY OF CONGRESS
CATALOGING-IN-PUBLICATION DATA

Clapp, Nicholas.
The road to Ubar : finding the Atlantis of the sands / Nicholas Clapp.
p. cm.
Includes bibliographical references (p.   ) and index.
ISBN 0-395-87596-X
1. Ubar (Extinct city).   2. Excavations (Archaeology) —
Oman — Ubar (Extinct city).   I. Title.
DS247.063C55 1998
939'.49—DC21   97-36640   CIP

Book design by Anne Chalmers
Type is Electra by Linotype-Hell
Line drawings and endpapers by Kristen Mellon

Printed in the United States of America

QUM 10 9 8 7 6 5 4 3

For Kay, Cristina, Jenny, and Wil

Contents

The Road to Ubar

Prologue

Boston, Massachusetts, February 1797 ... I T W A S S N O W I N G and well after dark when the wagon finally pulled up outside the book-shop on the corner of Proctor's Lane. Wil, the young proprietor, would have been waiting anxiously, stamping his feet to keep warm and every few minutes wiping the snowflakes from his spectacles. He helped unload the shipment of the books he'd had printed in New Hampshire and, back inside, hastened to inspect a copy. The sturdy little volume began with his friend Cooper's account of his trip to the continent and his discovery in a country inn of a French edition of the *Arabian Nights Entertainments*. Cooper wrote, "When I had finished reading the book, it struck my imagination, that those tales might be compared to a once rich and luxuriant garden, neglected and run to waste, where scarce any thing strikes the common ob-server but the weeds and briars, whilst the more penetrating eye of the experienced gardener discovers still remaining some of the most fragrant and delightful flowers."[1]

Wil paced back and forth in his tiny shop, leafing through the translation — the first in America — of the tales. It was a daring, even reckless thing that he had chosen to do. It was not so long ago that the Reverend Jonathan Edwards had deemed that the only fit reading was the Bible or commentaries on it. Works of the imagina-tion were the work of sinners, to be punished by an angry God. "That God holds you over the pit of hell," Edwards fulminated, "much as one holds a spider, abhors you, and is dreadfully provoked."[2]

Wil, though, thought he had sensed a recent change in public sentiment. People were tired of the dark cloud of Puritanism. The time was ripe, he thought, for the "most fragrant and delightful flowers" of the *Arabian Nights Entertainments*, which he had slyly retitled *The Oriental Moralist*, hoping that nobody would notice the rather striking absence of morality in these tales of evil magicians, flying horses, secret lovers, and haunted, lost cities.

Wil's *Oriental Moralist* included "The Petrified City," a tale told by Zobeide, an enterprising woman of Baghdad. Accompanied by two tiresome sisters, she sets out on a journey:

> We set sail with a fair wind, and soon got through the Persian gulph, and saw land on the twentieth day. It was a very high mountain, at the bottom of which we saw a great town. . . .
>
> I had not the patience to stay till my sisters were dressed to go along with me, but went ashore in the boat by myself, and made directly to the gate of the town. I saw there a great number of men upon guard, some sitting and others standing with sticks in their hands; and they had all such dreadful countenances that they frightened me; but perceiving they had not motion, nay not so much as with their eyes, I took courage and went nearer, and then found they were all turned into stones, all petrified.[3]

Zobeide, though frightened, is determined to find out what happened. Exploring the town's fantastical palace, she discovers it full of "infinite riches, diamonds as big as ostrich eggs." And she discovers a sole survivor, a man chanting the Koran, who relates: "It was about three years ago, that a thundering voice was suddenly and so distinctively heard throughout the whole city, that nobody could avoid hearing it. The words were these: 'Inhabitants, abandon your idolatry, and worship the only God that shews mercy.'"

It seems that the message was repeated for three years, until the "only God that shews mercy" apparently ran short of it, and at four o'clock in the morning petrified the entire population, with the exception of the fellow chanting the Koran, who joins Zobeide and her sisters as they leave the city. The tale now takes some curious turns. At sea, Zobeide's envious sisters push her and her new friend overboard. He drowns, she survives. For their treachery, the two sisters are turned into black dogs by a passing dragon. Back in Baghdad, Zobeide divides her time between enjoying her great riches (for she had gathered up a few souvenirs) and disciplining her two new black dogs. She allows that "since that time I have whipped them every night, though with regret."

The world of "The Petrified City" was a world unknown to puritanical and bleak New England. Prior to Wil's publication of *The Oriental Moralist*, American school geographies had had little to say of Arabia, other than that "the Arabs are an ignorant, savage and barbarous people. Those on the coast are *pirates*; those in the interior are *robbers*."[4] Yet in "The Petrified City," Zobeide is portrayed as smart, sensual, brave, and remarkably independent. And through her eyes we enter a world of exotic sights and sounds, of Oriental wisdom, of strange and mysterious happenings.

Zobeide's tale also happens to be the very first account printed in America of a city that time and again magically appears and disappears in the course of the thousand and one nights of the *Arabian Nights Entertainments*. The city is usually located in Arabia. Sometimes it is at the edge of the sea, but more often the traveler has to cross a forbidding mountain range and venture into a vast, sunscorched land. Sometimes the city has no name, but often it is called Iram. And, as we shall see, Iram is one and the same as a fabled land and city known as Ubar.

Ubar, rich beyond all measure. Ubar, for its sins, suddenly and dramatically destroyed by Allah.

Back in the winter of 1797, aspiring publisher Wil Clap could take pride in "The Petrified City" as one of the "most fragrant and delightful flowers" offered to his fellow New Englanders. Sadly, his offering was unrequited: *The Oriental Moralist* had only a single small printing. Though Wil survived by printing tracts and memoirs penned by his Puritan ancestors, he was eventually forced to close up shop and head west, then south, in search of business. On his way to New Orleans he died in his forty-eighth year, of unrecorded cause.

Wil meant well, and he made a remarkable unsung contribution. So it is fitting that this book is dedicated to a forefather I never knew: William T. Clap. His *Oriental Moralist* opened a door on a wondrous world. Nearly two hundred years later, my wife, Kay, and I and a hardy band of adventurers would have the good fortune, like Zobeide, to journey to a far land of the *Arabian Nights Entertainments* in search of its petrified city, in search of Ubar.

March 1997

N O T E : In this journey to unfamiliar places populated by unfamiliar people, both of the past and of the present, the reader may wish to consult Key Dates in the History of Ubar, page 275, and the Glossary of People and Places, page 277.

I

Myth

1

Unicorns

Over Iran, December 1980 . . . The small cargo plane flew on into a starry but moonless night.

"You cannot be up there," the voice crackled over the radio. "We are having a war here. You are not understanding? Yes?"

While the pilot worked the radio, the copilot tried to make some sense of the scattered lights below. Were they in southern Jordan or perhaps Saudi Arabia? No. It appeared that the aircraft had somehow strayed into Iran, which at the time was engaged in a heated war with Iraq.

"Okay, okay, okay. Got it," the pilot radioed back. With a sigh, he turned to the copilot. "We'll head west then? And sort things out." He paused. "Hopefully."

As the cargo plane banked, the flight engineer, wedged behind the copilot, checked his instruments — those that didn't have "INOP" stickers stuck to their faceplates. The oil leak seemed okay now, and the port engine wasn't overheating as long as they took it easy and held back on the throttle.

The journey had begun two days earlier in a winter storm that turned the San Diego Wild Animal Park into a sea of mud. In a driving rain, three of the zoo's rare Arabian oryxes — magnificent black and white animals with long, tapered horns — were patiently coaxed into a chute and loaded into large wooden crates. They were going home.

Once, great herds of oryxes had freely roamed Arabia. But in the early part of this century, the peninsula's bedouin began replacing their old flintlocks with accurate and deadly Martini-Henrys. A large oryx could feed a family for a month, and the hunt was exciting, a test of riding and marksmanship. Later, oil-rich princes joined the hunt, not on fiery Arab steeds but on military half-tracks fitted with heavy-caliber machine guns. For sport, not food, they would slaughter sixty or more animals in an afternoon. Until there were no more. By the early 1970s, the Arabian oryx was extinct in the wild.

Fortunately, a number of conservation groups had faced the reality that the animal was being wiped out in its native habitat and had initiated an innovative breeding program. Arabian oryxes in zoos were swapped back and forth so that a genetically sound "world herd" could be created. By 1980 there were enough animals in captivity that a few at a time could be returned to the wild.

On their journey home, San Diego's oryxes would have company: Dave Malone, a young zookeeper, and a documentary film crew, consisting of myself and my wife, Kay, cameraman Bert Van Munster, and soundman George Goen. As soon as the oryxes were secured in their crates, the clock began ticking, for it would be unwise to risk opening the crates to give the sharp-horned animals food or water. It was essential to get them to Arabia as quickly as possible.

The freeway north to Los Angeles was partially flooded and choked with traffic. The Wild Animal Park truck made it to Air France Cargo with not a moment to spare, and we and the oryxes were on our way to Paris. There we transferred to another cargo plane, flown by a pickup crew that normally worked for British Midlands. After nightfall they veered off course somewhere over eastern Turkey. The error was understandable. Of the crew, only the pilot had made the run before — once, ten years ago.

Now I was in a jump seat behind the pilot, except the pilot wasn't there. He was all but on hands and knees, puzzling with the rest of the crew over navigational charts spread out on the cockpit floor. Gazing into the night, I thought I saw something. A glint in the moonlight.

"By any chance could we have company up here, coming our way?"

"Doubt it. Not at this altitude."

"You're sure?"

"Actually, no."

The pilot swung up, peered ahead, didn't see anything. But his eyes weren't accustomed to the dark. He flipped on the plane's landing lights. And in response, coming at us, another set of landing lights lit up the sky, the beams diffused by the petro-haze that hovers miles high over Arabia. The two planes streaked past each other. Dave, who'd been back in the cargo hold checking on the oryxes, poked his head through the cockpit doorway.

"You guys okay?"

"Just fine," the pilot said.

And we were. A few minutes later the copilot spotted the burning flares marking Saudi Arabia's major north-south pipeline. "Flying the pipeline" took us to within an hour of our destination: Muscat, the capital of the Sultanate of Oman, where His Majesty Sultan Qaboos ibn Said had become intrigued by the plight of the oryx and had established a program to reintroduce the species into the wild.

At three A.M. we banked to the right just short of the silvery Arabian Sea and were on final approach to what the pilot was pretty sure was Muscat's Seeb Airport. We landed and barely had time for a catnap before three winged boxes emerged from a hangar and whirred toward us. They were Skyvans, small Irish-made military planes that could carry a small vehicle — or a crated oryx — and land it almost anywhere. The pilot in charge, Muldoon, Irish like his plane, supervised the loading with inordinate cheerfulness, considering the hour. Muldoon was a mercenary for Oman's fledgling air force. He was a *good* mercenary, he took pains to explain, busy with worthy missions (food drops, medical flights, and so on) in a time of peace.

We boarded Muldoon's plane. He flashed a thumbs-up and hit the throttle. Despite being loaded down with oryxes and fifty-five-gallon drums of fuel for the return flight, our three planes were quickly airborne. We circled over the sea to gain altitude and greeted the dawn as we headed toward the Jebel Akdar, the rugged "Green Mountains" that rise abruptly from Oman's coast. The greenery at first was limited to tiny terraced cornfields and vineyards. But then we flew into a long, winding valley and over grove after grove of palm trees.

Beside me, Kay had her face pressed to the window, taking all this in. Neither of us had ever been east of Europe, much less flown a barely charted desert in a tiny, mercenary-piloted plane. This didn't faze Kay a bit; she loved it. In everyday life, though, some things did faze her. Raised in the South, she could become distraught upon discovering that her navy shoes didn't match her new navy skirt or, worse yet, that her hair had become "a mop, with simply nothing to be done about it." Big things, like a crazed teenager trying to knife her or an international dope dealer threatening to have her "disappeared," didn't bother her at all. Our documentary filmmaking jaunts were breaks from her job as an in-the-trenches federal probation and parole officer. I remember her coming home one day all black and blue.

"Mom, what happened to you?" inquired first-born daughter Cristina.

"More aikido training with the FBI," she said nonchalantly. "This morning it was how to slow bad people down by, um, doing things to their kneecaps."

Always chipper, immensely capable, Kay is a good partner in strange places. We unbuckled our seat belts and squeezed by a crated oryx for a view from the cockpit. "The way to the interior," Muldoon the (beneficent) mercenary gestured, as our three Skyvans buzzed a crumbling old watchtower and cleared a narrow pass.

Ahead now was a vast, rocky plain dotted with mud-brick villages. But soon the villages were behind us, all but one, set in a lonely cluster of palms. "Adam, the oasis of Adam," Muldoon said, then mused, "Suppose that's where he and the missus got the gate?"

The oasis was a last landmark. Oman's interior, desolate and featureless, rolled off to the horizon. We droned on for an hour. The Skyvan couldn't go very fast and, with no pressurization, had to stay under 5,000 feet.

Ahead, fingers of red sand reached out for us. "The Rub' al-Khali?" I ventured, surely mispronouncing the Arabic for the Empty Quarter.

"If you want it to be," Muldoon replied. "Who knows where it begins?"

The Empty Quarter is the great sand sea of Arabia, the largest sand mass on earth. Following the fingers of sand to the horizon, Kay and I could see — or thought we could — distant dunes, dancing through the heat waves. And then the fingers of sand were gone, left

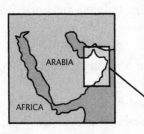
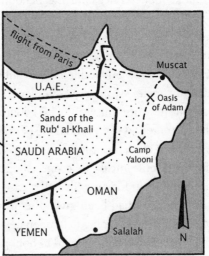

Land of the oryx

behind. Muldoon squinted ahead and began his descent to Camp Yalooni. Beyond the reach of roads, with scant vegetation and no water (the nearest well was eighty miles away), it was the ideal place to release our oryxes, as far as possible from harm's way. A scattering of specks became a cluster of small prefab buildings and a water truck. No airstrip. Muldoon circled once, slowed till the plane's stall alarm went off, and hit the rocky terrain with a bump and a crunch.

By now the oryxes had been in their crates for just over sixty hours.

Clambering out of the Skyvans, we were greeted by Mark and Susan Stanley-Price, the personable wildlife biologists in charge of Camp Yalooni. Behind them, running across the desert, came a band of bedouin, shouting and waving rifles. Members of the Harasis tribe, they were garbed in turbans and long robes. Wickedly curved daggers were tucked into their belts, and state-of-the-art Motorola walkie-talkies hung from their shoulders. They were to be the oryxes' gamekeepers.

Mark Stanley-Price and the bedouin shouldered the first crate from the plane and carried it to the edge of a nearby fenced enclosure, the holding area for the animals until they were turned loose in the desert. Dave Malone scrambled up onto the crate and unlatched its sliding door. Mark nodded, Dave pulled up, and the first oryx flew out of the crate. We cheered. He slowed to a trot and circled, not the least bit the worse for wear. The bedouin broke into a tribal chant. The two other oryxes repeated the performance.

In honor of the occasion — or so we assumed — the Harasis prepared a favorite meal: Take one whole, tokenly eviscerated sheep, add rice. Cook. Flavor with half a case of La Ranchita taco sauce. From the day I had been given the okay to go to Arabia, I dreaded what I was sure was going to happen next. The sheep's eyeballs, I had read, were traditionally offered to honored guests. Kay had a plan, at least for herself. She would lower her eyes, and murmur words never to be breathed outside of Arabia: "Oh, how kind, but I'm not worthy, for I'm just a woman." Inevitably, an orb (perhaps two?) would be in my court. Were they viscous and slimy? Crunchy?

I was relieved when, apparently unaware of this tradition, the Harasis bedouin unceremoniously dug in, the dread orbs disappearing in a melee of hungry hands. The bedouin were fast eaters — to avoid surprise attack, it's been said, but also, I suspect, to get the best parts and leave the gristle to the poky. When they rose from the feast, they were in an exceptionally good mood. They unsheathed their daggers and broke into a wild impromptu dance that somehow turned to terrorizing zookeeper Dave. He was a good, if nervous sport. As knives swiped within an inch of his nose, he pleaded to little avail, "Why me? I'm from New York."

"They're a little cranked up today," observed Mark Stanley-Price.

"It's a big event, the oryxes coming in," I added.

"The oryxes? Oh my, no, dear me. These chaps came back this morning from raiding their rivals, the next tribe off into the interior. Dynamited their best well, I hear."

"Oh . . ." And I got a glimmer that even if ecology was not a major part of the Harasis ethic, it wouldn't be a very good idea to lay a hand on the oryxes they were now charged to protect.

Late that afternoon, when Camp Yalooni's drab plain turned fleetingly golden, Kay and I walked over to visit the oryxes. And we saw that myths could be real. Here it was the myth of the unicorn.

Though unicorns appear in Persian and biblical chronicles, their heyday was in medieval Europe. It has been suggested that a lone traveler to Arabia spied an oryx in profile, with one horn masking the other. On his return home, he entranced his friends and ultimately all of Europe with the vision of a magnificent one-horned creature. This seems unlikely, though, for even minimal and distant oryx-watching will be rewarded by a flick of the head and a view of the animal's two long spiraled horns. It is much more plausible that a single horn (minus oryx) made its way to Europe, and a horselike creature was dreamed up to go with it.

Either way, the Arabian oryx appears to have been the inspiration for the legendary unicorn. As described in a medieval book of beasts, he has "one horn in the middle of his forehead, and no hunter can catch him. . . . He is very swift because neither Principalities, nor Powers, nor Thrones, nor Dominations could keep up with him, nor could Hell maintain him." Only a fair virgin could approach a unicorn and hear him say: "Learn from me because I am mild and lowly of heart."[1]

Two of our oryxes were quietly foraging. The third was silhouetted against the setting sun. At a glance, the animals looked too delicate, too ethereal to survive in a land as harsh as this. They were certainly graceful, but they were also incredibly rugged. Sixty hours in a box was nothing. They could go days — a lifetime, if need be — without water, getting all the moisture they needed from scant forage. Comfortable in searing days and freezing nights, the oryx survived as if by magic. It was hard to imagine this lifeless landscape nurturing a mouse or a bird, but nevertheless . . .

This was where unicorns lived.

2

The Sands of Their Desire

"THE ODYSSEY OF THE ORYX" proved to be a popular segment of the television series *Amazing Animals*. Bert, George, and I were now dispatched to do a series of domestic stories, some more edifying than others. We covered Bart the Kodiak Bear and Buster the Wonder Dog. "The wonder of that dog Buster," noted cameraman Bert, as Buster demurred at walking his tightrope, "is what that dog doesn't, can't, or won't do." Yet with prompting and patience, Buster finally teetered across his tightrope, jumped through a flaming hoop, and dove from the Malibu Pier, demonstrating his prowess should he ever be called upon to aid a sinking swimmer.

Over the next few months, Kay and I thought often about Arabia. As was our custom, we dined frequently at the El Coyote Spanish Cafe, known for its margaritas and motherly waitresses, got up in beehive hairdos and fuchsia hoop skirts ample enough to conceal steam tables. A conversation between Kay and me would go: "How about trying a number-six special for a change?"

"You can. I'm sticking with a number one. Nice to see all the regulars." (We had passed Ricardo Montalban on our way in and were seated across from the Twins, two elderly, nattily attired gentlemen who dined at El Coyote every single night.) Then, with no transition, "How do you think the oryxes are doing?"

We talked of them and of their keepers, the Stanley-Prices. We remembered that Camp Yalooni had been buggy. The next day Kay purchased a case of Cutter insect repellent and sent it off to them. A week or two later, back at El Coyote, we further wondered: if the oryxes could survive in the interior of Oman (which they did, spectacularly), what other wonders might the Arabian desert hold? What would it be like to venture into the Rubʿ al-Khali?

"A reason," Kay said. "We need a reason, a way to go back."

We read up on the Arabian peninsula, its natural history, its geography, its exploration. Within walking distance of the *Amazing Animals* editing rooms, I discovered Hyman and Sons, a bookstore specializing in Egyptology with a scattering of books on Arabia. I quickly came to appreciate how fortunate we had been to set foot in the peninsula's interior, to even glimpse the sands of the Rubʿ al-Khali.

For centuries Arabia had been terra incognita, a mysterious medieval land out-of-bounds to Western exploration. What little had been written discouraged outsiders. In the 1400s Sir John Mandeville characterized the Arabian bedouin as "right foul folk and cruel and of evil kind." A 1612 account elaborated: "The people generally are addicted to Theft, Rapine, and Robberies; hating all Sciences Mechanicall or Civill, they are commonly all . . . scelerate and seditious, of coulour Tauny, boasting much of their triball Antiquity, and noble Gentry."[1]

But then, beginning in the early 1800s, a succession of adventurers penetrated Arabia, concealing their identities by donning native costumes and creating elaborate cover stories. We don't know how Ulrich Seetzen, a Swiss biblical scholar, disguised himself, but whatever it was, it didn't work. At some point in his 1806 journey, he was set upon and murdered by fierce bedouin, their suspicions possibly aroused by his interest in ancient ruins. Wishing to avoid a similar fate, his countryman Johann Burckhardt darkened his face and hands with the juice of the betel nut and adopted the guise of a wandering physician from India. But when he opened his mouth, bedouin eyes narrowed. His Arabic was strangely accented. Of course it was, he responded, and unleashed a volley of guttural German. Perhaps the bedouin were unaware, he glibly explained, but this was how the Muslim faithful conversed in India. It was only in Cairo — after discovering the lost city of Petra and even entering forbidden Mecca — that Burckhardt's ruses reportedly failed, and he was poisoned or beheaded. Or he may have died of fevers contracted in his Arabian journeys; accounts vary.

Others, against considerable odds, lived to tell their tales. In *Incidents of Travel in Egypt, Arabia Petraea and the Holy Land* (1837) the American John Lloyd Stephens gave a hilarious account (not so hilarious at the time) of the down side of his elaborate getup. His magnificent turban, long red silk gown, and curly-toed yellow satin Turkish slippers were distinct liabilities when, his infidel identity suspected, he had occasion to flee, on foot over rocky terrain, a band of irate bedouin. He "dashed down the mountain with a speed that only fear could give. If there was a question between scramble and jump, we gave the jump."[2]

I really admired Stephens. Imagine a New York lawyer in failing health who was advised by his physician to seek a cure in travel and rest — and chose to venture across the Sinai and the deserts beyond, where no American had ever set foot. Certainly the desert Arabs had a weakness for brigandage, Stephens noted, but also they valued poetry, and they had a streak of chivalry that, if need be, dictated sharing their waterskin with their worst enemy. They believed that all others, like themselves, were guests of God in the wilderness and should not be denied God's gifts of sustenance and shelter.

What comes across in this and other accounts is that the Arabs of the desert were perhaps excitable and hot-tempered yet, surprisingly, not that intolerant of adherents of Western religions. They may have been riled more by deception than by infidelity. Englishman Gifford Palgrave journeyed deep into Arabia, freely admitting that he was a simple "Jewish Jesuit."[3] And his countryman Charles Doughty persistently chided his Arab hosts for their lack of Christian charity (which they confirmed by throwing him in jail with predictable regularity). Even so, they allowed him to travel back and forth along the northern reaches of the Rub' al-Khali.

Anxious to obtain a copy of Doughty's masterpiece, *Travels in Arabia Deserta* (1868), I stopped by Hyman and Sons bookshop on a Saturday morning. A bell tinkled as I stepped inside, and I found myself the sole customer amid a jumble of books shelved, stacked on the floor, piled on tables. One table had been cleared off and was set with a pitcher of milk, glasses, and a plate of cookies. The door to a back room swung open.

"Go ahead. Somebody's got to eat them." It was Virginia. Though I never saw anything of Hyman, much less his Sons, I had gotten to know and respect Dr. Virginia Blackburn, who was clearly in charge here. She was in her late sixties, and if there was a word to describe her it was "formidable." "Gruff" also comes to mind.

"Charles Doughty," I inquired. "*Travels in Arabia Deserta.* Have it by any chance?"

"No," said Virginia.

"Well, then, maybe . . ." I started to say, wondering if the book could be tracked down.

"No," repeated Virginia, peering over her steel spectacles. "You wouldn't like it. Not at all." Then, hesitating not a second, she rounded the counter, marched across the shop, ran her finger across a shelf, and removed a modest worn blue volume. "Read this."

"Well, no, what I really . . ."

"Blowhard. Full of himself, I'll tell you," Virginia proclaimed, coming down pretty hard on poor Doughty. "Tried to reinvent the English language, making it courtly like classical Arabic." By way of contrast she held out the blue volume. On its cover was a lone, faded golden figure riding a camel and the title *Arabia Felix.* "Read this."

"Well, well I'm not sure . . . but okay." Having sampled the cookies, it seemed impolite to leave empty-handed.

"Good. You'll like it. He had red hair, like you," she said, referring to the author of the book, Bertram Thomas.

Bertram Thomas, I soon learned, was an amazing, tenacious, extremely likable fellow. But for his refreshing lack of self-promotion, he would be counted among the greatest of Arabian explorers. He writes nothing of his life prior to the day in the late 1920s that he set foot in Arabia Felix ("Happy Arabia," the old Roman name for southern Arabia), where he had accepted an offer to become the wazir, or financial adviser, to the sultan of Muscat and Oman. It was hardship duty. For the better part of the year, the small coastal city of Muscat baked and sweltered like no other place on earth. Nighttime temperatures refused to drop, as they did inland in the desert. At two o'clock in the morning, the temperature was as high as 124 degrees Fahrenheit, the humidity an unbearable 100 percent.

Thomas's Muscat days were spent managing, as best he could, the accounts of a court and country that exported only firewood, rotting sardines (for fertilizer), and slaves. Every day at sunset, a cannon was fired, and the gates of the walled medieval city swung shut. Until dawn the next day, everyone was confined to their dwellings, with the exception of the chosen few to whom the sultan had issued a special identifying lantern.

As wazir, Thomas had a lantern, and at night he could roam the fitfully sleeping city. He could climb its medieval mud-brick walls and gaze away to the north. Beyond the coastal mountains, he knew, the moon that shone on Muscat shone also on the dunes of the Rub' al-Khali. The desert was the secret reason Thomas had come here: he had long wanted to be the first Westerner not only to venture into its heart but to actually cross it — even though T. E. Lawrence (Lawrence of Arabia) had determined that "nothing but an airship can do it."[4]

In his time in Muscat, Bertram Thomas changed. A photograph taken early on shows him posed in front of a crumbling doorway, incongruously dressed in a wool suit and felt hat. But soon the hat was replaced by a jaunty turban, the somber suit by a flowing robe. He raised a beard and carried a camel stick. And Thomas found excuses to venture away from Muscat. On his travels he got along famously with desert tribesmen. In their company he rode by camel to the edge of the Rub' al-Khali and saw for himself that (as a bedouin saying went), "Where there is no water, that's the Empty Quarter; no man goes thither."

Back at the sultan's court, Thomas was asked a question:

"Why aren't you married, O Wazir?" was fired off at me by an un-comprehending Arab.

I expatiated on the difficulties under which a Christian labored, especially one serving in the East, and pointed to the comforting doctrine that for a man it was never too late.

"Ah!" said the Sultan, knowing of my secretly cherished desire, "quite right. *Insha'allah*, I will help to marry you one of these days to that which is near to your heart — *Rub' al-Khali. Insha'allah!*"

"A virgin indeed," quoth Khan Bahadur, his private secretary.

"*Amin!*" I muttered to myself. "So may it be."[5]

But to Bertram Thomas's increasing dismay, the sultan was reluctant to let his wazir go roaming north across the great desert. And a rival now threatened Thomas's dream. In Riyadh, in the kingdom of Saudi Arabia, Harry St. John Philby, a flamboyant Arabist late of the British Foreign Service, was poised to attempt a crossing of the Rub' al-Khali in the opposite direction, north to south.[6] The two men knew each other. Earlier in the decade, Thomas had spent some time working for the British Foreign Service in Transjordan. Harry Philby had been his superior and had advised Thomas to take the Muscat and Oman assignment, confiding that Muscat was "the best starting point for crossing the Empty Quarter." There's a hint of duplicity here. Did Philby make the suggestion out of the goodness of his heart? Or did he see this as an opportunity to get Thomas out of the way, to put him in the clutches of a possessive and paranoid sultan, so that he, Philby, could claim exploration's last great prize?

The rivalry was heightened by a strange shared vision. Both Thomas and Philby saw the Rub' al-Khali as a beckoning yet veiled virgin. Thomas called the Rub' al-Khali "the sands of my desire." Philby called the same sands "the bride of my constant desire."[7] But, though there were two suitors, there could be only one husband.

On an October night in 1930, unbeknownst to his sultan, Bertram Thomas stole away from Muscat with the anticipation that "tomorrow, the news of my disappearance would startle the bazaar and a variety of fates would doubtless be invented for me by imaginations of oriental fertility."[8] He rowed to a rendezvous with a passing oil tanker and "ere four bells had struck" was on his way by sea to the southern Omani town of Salalah. There he would find guides and outfit his expedition.

Thomas liked Salalah. The place had a cheerful African air. He was particularly amused by the unusual fate of the town's black slaves. Their Arab masters led grim, obsessive lives, forever worrying over the shame that would fall upon them if any of their veiled, sequestered wives or daughters took a wayward turn. By contrast, a household's slaves enjoyed a happy-go-lucky freedom unimaginable to their masters. And, dancing and singing in the dirt streets of Salalah, they saw Thomas off on December 10, 1930, as he headed north with a train of fifteen camels and a rascally band of bedouin of uncertain repute. Salalah's *wali*, or mayor, had warned Thomas, "If there is a thing they do better than lying, it is stealing."

Far to the north, in Riyadh, Harry Philby was informed that this year, at least, he would not be granted permission to travel south across the Rub' al-Khali. If he would like, the following year he could again petition the king, 'Abdul 'Aziz ibn Sa'ud.

Crossing the Dhofar Mountains, which abruptly rise beyond Salalah, Thomas displayed a fine eye for geographic and ethnographic detail. In *Arabia Felix* he noted that the mountain tribesmen spoke a strange, non-Arabic language and considered themselves descendants of a mythical race known as "the People of 'Ad." He witnessed blood sacrifice and an exorcism performed with frankincense and fire. All the while, he charted his progress with an accuracy that is amazing considering that he had to make his navigational sightings in secret, "lest I be suspected of magic or worse."[9]

Beyond the mountains lay a "barren plain, sun-baked and filmy with mirages," desolate but for occasional sightings of Arabian oryxes, running wild and free. Five waterless days took Thomas to the remote waterhole of Shisur, the site of an abandoned "rude fort," a last, forlorn trace of civilization. A day beyond Shisur, Thomas sighted "the sands of my desire." But his little band did not immediately plunge into the Rub' al-Khali. They skirted its southern edge, on the lookout for the first of a series of wells they hoped would see them safely across the dunes.

The seventh waterless day out of Shisur began like the others. Thomas wrote:

> Our morning start was sluggish. We straggled because of the cold and the hunger and the many transverse sand ridges, and straggling camels mean a slow caravan. An hour's march brought us to a wide depression. . . .
>
> Suddenly the Arabs, who were always childishly anxious to draw attention to anything they thought would interest me, pointed to the ground. "Look, Sahib," they cried. "There is the road to Ubar."
>
> "Ubar?" I wondered.
>
> "It was a great city, our fathers have told us, that existed of old; a city rich in treasure, with date gardens and a fort of red silver. (Gold?) It now lies buried beneath the sands in the Ramlat Shu'ait, some few days to the north."
>
> Other Arabs on my previous journeys had told me of Ubar, the Atlantis of the sands, but none could say where it lay. All thought of it had been banished from my mind when my companions cried

their news and pointed to the well-worn tracks, about a hundred yards in cross section, graven in the plain. They bore 325°, approximately lat. 18°45′N., long. 52°30′E. on the verge of the sands.[10]

On his remarkably accurate map of Arabia, prepared for the Royal Geographic Society, that is where Bertram Thomas noted "the road to Ubar."

It was an unexpected, exciting discovery, and Thomas must have been tempted to follow the impressive road to the fabled city. But a side trip at this point would have depleted his waterskins and jeopardized his dream of reaching the desert's far side. Passing the road to Ubar by, Thomas embraced the Rub' al-Khali, whose virgin landscape he found less an enchanting bride and more "a hungry void and an abode of death." Great was his relief when, ninety-five days after leaving Salalah and the Arabian Sea, he came in sight of the town of Doha and the Persian Gulf. He'd done it! The Rub' al-Khali was his.

In Riyadh, the news so enraged and disheartened Harry Philby that

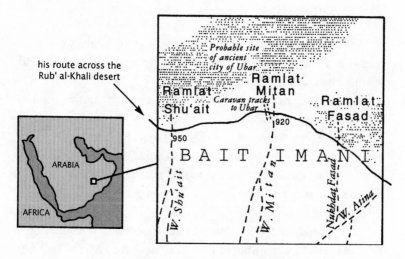

Detail of Bertram Thomas's map of Arabia

he shut himself indoors and refused to come out for a week. When he did, he made no effort to conceal his feeling that he had been betrayed by his friend and protégé. Quoting a verse of Arabic poetry, he expressed his hurt: "'Twas I that learn'd him in the archer's art; / At me, his hand grown strong, he launched his dart."[11]

Philby's petulance aside, Thomas's achievement was greeted with acclaim. T. E. Lawrence called it "the finest thing in Arabian exploration." He wrote, "Bertram Thomas has just crossed the Empty Quarter, that great desert of southern Arabia. It remained the only unknown quarter of the world, and it is the end of the history of exploration."[12]

Or was it? Thomas had crossed the desert but had by no means thoroughly explored it. His very journey had a tantalizing loose end: a mysterious byway, a road leading to a lost city of the sands. Granted that his knowledge of Ubar came from his not-known-for-their-truthfulness bedouin companions. ("If there's anything they do better than lying, it's stealing.") Still, there was the fact of the road itself, witnessed by a keen observer, a man not to be doubted. And all roads lead somewhere.

Fifty years later, half a world away, over a #1 and #6 at the El Coyote Spanish Cafe, I wondered, and Kay did too: Might we have a reason to return to Arabia? Had Bertram Thomas gone back? Had anyone else taken up the search for Ubar?

Yes, they had. Revisiting Hyman and Sons and haunting the DS 200–250 history stacks at UCLA's University Research Library, I found that Thomas's report had made Ubar a touchstone of Arabian exploration. The famous, the foolhardy, and at least one out-and-out charlatan had taken up the quest to find the lost city. As for Bertram Thomas, he never returned to Arabia, though there was a wisp of evidence that he visited Mecca. In a trunk of his memorabilia, which is in the custody of the Institute of Oriental Studies in Cambridge, England, there's a snapshot of him in front of the "Mecca Post Office." But why is the sign in English, not Arabic? And what is that over Thomas's shoulder, barely visible through the post office window? With a magnifying glass, I could make out *Coll . . . Colliers*, a popular American magazine of the 1930s. Thomas was not in Mecca, holy city of Arabia, but in Mecca, a sleepy desert farm town in southeastern California, where he had stopped on a lecture tour following the American publication of his *Arabia Felix*.

Back in Arabia, Harry Philby, at last and too late, was finally given permission to venture into the Rub' al-Khali. He did so, for he saw a chance to even the score with his protégé-turned-rival. *He* would discover Ubar. His success was, in fact, assured.

Some years before, in a rare year of rain, bedouin dwelling on the northern fringes of the sands had followed their flocks deep into the Rub' al-Khali and had happened upon Wabar (as Ubar was also known). They now agreed to lead Philby back to its ruins, the ruins of a city so rich that pearls still lay scattered in the sand. They described as well a large half-buried camel forged of iron.

In March of 1932, Philby rode south from Riyadh and into the Rub' al-Khali. Forgotten, for the time being, was his grudge against Thomas. He and his bedouin were in high spirits. His companions sang:

> Hear then the words of 'Ad [Ubar's first king], Kin'ad
>   his son:
> Behold my castled-town, Aubar [Ubar] yclept!
> Full ninety steeds within its stalls I kept,
> To hunt the quarry, small and great, upon;
>
> And ninety eunuchs tended me within its walls
> Served in resplendent robes from north and east;
> And ninety concubines, of comely breast
> And rounded hips, amused me in its halls.
>
> Now all is gone, all this with that, and never
> Can aught repair the wreck — no hope for ever![13]

In the late afternoon of the nineteenth day of his journey, Philby drew in his breath. Ahead, rising from the sands, were the blackened walls of Wabar, scorched by the fires of its destruction. His bedouin, heartened by the promise of fortune, cheered wildly. Philby, heartened by the promise of fame, raced his camel across the dunes.

His hopes were suddenly and completely dashed. For instead of Wabar, he came upon a circular crater in the sands, "a work of God not man." "I knew not whether to laugh or cry, but I was strangely fascinated by a scene that had shattered the dream of years. So that was Wabar! A volcano in the desert! and on it built the story of a city destroyed by fire from heaven for the sins of its King."

As if Wabar's mythical king were somehow responsible for his bitter disappointment, Philby railed on: "He had waxed wanton with his horses and eunuchs and concubines in an earthly paradise until the wrath came upon him with the west wind and reduced the scene of his riotous pleasures to ashes and desolation!"

But what of the pearls to be found here, scattered in the sands? Philby watched as the bedouin "burrowed for treasure, and took small shiny black pellets to be the pearls of 'Ad's ladies blackened in the conflagration that had consumed them with their lord." In reality, worthless globules of crystallized glass ran through their fingers.

And what of the reported great iron camel? The bedouin scoured the site; it was nowhere to be found. They confessed they had never actually seen it, only heard about it from their fathers' fathers. Philby later learned that the great iron camel was in the basement of the British Museum. It seems that in the spring of 1863, a band of Rub' al-Khali bedouin, in the midst of a thunderstorm, had seen a meteorite fall from the sky. They found a large fragment of it — a fragment that resembled a camel — and carted it off. How it found its way to the British Museum is a mystery, but there it was stabled.

What Philby failed to understand was that he had, in fact, made a significant geological discovery. What he first took to be a desert volcano was in reality the impact crater of a meteorite; at the time only four or five had been found worldwide. Yet the discovery of the "Wabar crater" (as it is marked on modern maps) was of little consolation to Harry St. John Philby. On a searing desert day in 1932, his dreams proved only dreams, and he thereafter scoffed at the idea that there ever was such a place as Wabar or Ubar.

In England, T. E. Lawrence wasn't so sure. Lawrence's life had come to a strange pass. Assuming the name of T. E. Shaw, he had sought anonymity and obscurity among the rhododendrons of Dorset. Living in semiretirement in his Cloud Cottage, he sought to distance himself from his role as leader of World War I's celebrated Arab revolt. He felt that when all was said and done, he had betrayed the Arabs. He lamented his "mantle of fraud in the east," yet he considered returning to Arabia. The mantle "might be fraud or it might be farce: No one should say I could not play it."[14]

If he revisited the scene of his exploits, Lawrence indicated, it might well be to search for archaeological remains in the Rub' al-Khali. His friend Bertram Thomas had proven that this forbidden region could be penetrated and had brought back evidence that an "Atlantis of the Sands" — Ubar — lay hidden in its heart. Lawrence told an acquaintance, "I am convinced that the remains of an ancient Arab civilization are to be found in that desert. I have been told by the Arabs that the ruined castles of the great King 'Ad, son of Kin'ad, have been seen in the region of Wabar. There is always some substance to these Arab tales."[15]

Lawrence was certainly the most likely candidate to take up the search for Ubar. An erudite Arabist, he was also a trained archaeologist with field experience in Syria. He had a deep, near-mystical feeling for Arabian lore; indeed, the white-robed, blond figure known to the tribes as "al-Aurens" ("Lawrence") had become part of it. They would surely welcome his return.

In the early morning of April 2, 1935, was Lawrence thinking about how he had betrayed his Arab friends and followers? Was he dreaming of castles in the sand? Or was he simply caught up in the thrill of speeding down a deserted Dorset lane on his powerful Morris motorcycle? Suddenly, just ahead, two boys dodged onto the road. Lawrence swerved, lost control, and crashed. He hung on in a coma for a few days, then died. In Arabia the tribes would never again take up the cry of "al-Aurens! al-Aurens!"

A few years later, the world was caught up in another great war, and in its course Arabist-adventurer Wilfred Thesiger was dispatched to southern Arabia by the British Foreign Service. His mission was to find the desert breeding sites of the locusts that periodically swarmed out of the peninsula and destroyed the crops of Africa. This task, he found, provided a ready excuse to explore and write about the desert wilderness of the Dhofar region of Oman. His evocative, austere book, *Arabian Sands*, opens with the lines: "A cloud gathers, the rain falls, men live; the cloud disperses without rain, and men and animals die."

But, Thesiger readily acknowledged, "this cruel land can cast a spell that no temperate clime can match." At first Ubar didn't appear to be part of that spell; the spell "this cruel land can cast" had more to do with the privations of long, torturous marches with the bedouin, privations he seemed, a bit perversely, to enjoy. In *Arabian Sands*, Thesiger mentioned Ubar only in passing, as the sort of thing the bedouin argued about over their campfires.[16]

Thesiger said nothing of actually looking for Ubar. Yet a map included with his 1946 report to the Royal Geographic Society tells a different story. The routes of his major desert journeys are marked with dotted lines, one of which traces a journey that he wrote not a word about. Through waterless terrain, this dotted line takes him north to latitude 18°45′N, longitude 52°30′E, the very position where, twenty years before, Bertram Thomas had encountered his road to Ubar. Thesiger apparently ventured no further, instead retreating the way he had come. Had he sought Ubar but been forced to turn back, perhaps for lack of water? It appeared so.[17]

While Thesiger and his bedouin companions continued to roam Dhofar, yet another quest for Ubar began. On a moonless night in 1945, over the crackle of his campfire, Thesiger just might have heard a Royal Air Force Lodestar winging overhead. The plane had left Salalah bound for Muscat. The crew thought they were following the Arabian coast, but in fact they had made a navigational error and were on a false heading that took them inland over the desert.

As the Lodestar was struck by the first light of dawn, the pilot looked down, expecting to see the dark waters of the Arabian Sea. Instead he saw, from horizon to horizon, the sand sea of the Rub' al-Khali. The plane's crew scrambled to get a fix on their position and calculate their fuel reserves. They were well aware of the many World War II aircraft that had gone missing in the desert, a fate brought home as they flew over the wreckage of two Italian planes that had attempted long-distance strikes on the oil fields of eastern Arabia.

The Lodestar was lucky. Shooting the sun, the pilot got his bearings. They had just enough fuel to make it to an RAF base in the emirate of Sharjah.

On their heading to Sharjah, the plane's crew looked ahead to see a bowl-shaped mesa rising from the dunes. Seen from above, the bowl sheltered walls, towers . . . a city, a lost city! The aviators plotted its position. It would be easy to find: the ruin-crowned mesa was within sight of a known desert landmark, the palms of the well of Lihan.

Stationed at the RAF's Sharjah base, an airman by the name of Raymond O'Shea was entranced by the report filed by the pilot of the errant Lodestar. O'Shea had a two-week leave coming up, enough time for him and his mates to requisition a four-wheel-drive truck and travel overland to the site he believed to be "Qidan, the lost city of the people of 'Ad."[18] The people of 'Ad, I knew, were the people of

Ubar. Qidan, then, would be none other than Ubar. It fit. The site was in the area that had cruelly disappointed Harry St. John Philby. Had Philby given up too readily?

Crossing the desert by truck and then by camel, O'Shea found his way to Qidan with relative ease. It was, he claimed, an impressive place. The city's four-foot-thick walls enclosed acres of ruined buildings, and two forty-foot watchtowers were still standing. And there the search for Ubar might have ended. Describing a site difficult for anyone else to check out, O'Shea might have been lionized by the Royal Geographic Society for finding the lost city. But he made a mistake. In the account he wrote of his exploits, *The Sand Kings of Oman*, he featured a photograph of what he claimed was Qidan.

To archaeologists, the structures in O'Shea's photograph were timeworn but not ancient. To James Morris, a travel writer who had sojourned in Muscat and Oman, the structures were all too familiar; he had passed them many times. Morris wrote, "I realized with a start that Mr. O'Shea's illustration of his legendary city, which I had studied with respectful interest, indisputably showed our well-known road into Muscat. There is something almost Oriental about the glorious effrontery of the Irish."[19]

In all fairness to Mr. O'Shea, there may have been some truth to his story. It may be that he simply got carried away promoting it and couldn't resist the temptation to caption a photograph of Muscat as "Qidan." His mysterious mesa might well have been a desert outpost dating to the 1700s or even earlier.[20]

A few years after World War II, a dispute over drums of liquid rubber latex gave rise to a last devil-may-care attempt to find Ubar. The search was led by Wendell Phillips, a young American who had initially gone to Arabia to excavate Ma'rib, a site in Yemen he believed to be the royal city of the queen of Sheba. Phillips cut quite a figure. His signature costume included a checkered Arab kaffiyeh wrapped about his head, aviator sunglasses, a pearl-handled Colt .45

in a tooled leather holster slung low around his waist, and cowboy boots. He didn't object when reporters called him "Phillips of Arabia." Though only in his twenties, with modest academic credentials (a B.A. in paleontology from Berkeley), he managed to assemble an impressive staff of experts for a head-on assault on the antiquities of southern Arabia.

At Ma'rib, Phillips's team started by clearing away the sands that had swept in from the Rub' al-Khali and nearly buried the Mahram Bilqis, the reputed moon temple of the queen of Sheba.[21] Almost immediately they found finely chiseled inscriptions. The expedition's epigrapher, or inscription specialist, the Jesuit academic Albert Jamme, "almost trampled over the rest of us to get close enough to read them."[22] Beside himself with excitement, Father Jamme set to copying the inscriptions by lathering them with latex, then peeling off "squeezes" that three-dimensionally reproduced the elegant letters of the Sabaean (Sheban) alphabet.

The local sheik, who provided laborers for the excavation, was puzzled. What could explain the priest's elation? He reasoned that it must have something to do with treasure; only treasure could make a man so happy. As the good Jesuit's latex squeezes seemed a particular source of joy, the sheik began demanding — and receiving — duplicate copies of each new inscription.

But contemplating his latex squeezes brought the sheik neither happiness nor wealth. Tiring of rows of incomprehensible letters, he fixed on the idea of the latex itself. Though he couldn't quite understand why, latex equaled treasure, and it was only right that he should have his share of the expedition's remaining fifty-five-gallon drums of the liquid rubber. Phillips resisted. The sheik was furious. Things got ugly. A tense cable from Ma'rib read: "JAMME NOW HELD VIRTUAL PRISONER MARIB STOP QADI ZEID INAN DEMANDS JAMMES RUBBER LATEX COPIES OF INSCRIPTIONS STOP ALL ARCHAEOLOGICAL SPECI-

MENS LOCKED UP STOP GOVERNOR HOLDS KEY STOP FEAR SITUATION GETTING OUT OF HAND." Even though Father Jamme said "his latex squeezes meant more to him than life itself," the Phillips expedition elected to pack up and flee across the desert.

Escaping by boat from Yemen, Wendell Phillips could have headed home, but instead he followed the coast of the Arabian Sea east to the Dhofar region of Oman and, among other enterprises, took up the search for Ubar. At the wheel of a stake truck borrowed from the wali of Dhofar, Phillips made his way to the edge of the Rub' al-Khali. Passing a lone bedouin, he asked directions: "When I enquired if he knew the location of Ubar he shouted into my ear *faqat ash-shaitan ya'rif*, 'Only the devil knows.' I shouted back *wallahi sahih*, 'True, by God.'"[23]

Phillips suspected the bedouin might be right, for, hard as he looked, the great road reported by Bertram Thomas was nowhere to be found. In danger of running out of gas, he and his crew gave up. In their retreat they fortunately chose a different way than they had come.

> Regretfully we had turned back, heading east just south of the great dunes, when suddenly Charlie exclaimed, "There are the tracks!" It was California Charlie, not my desert-bred guide, who located these rows of parallel tracks incised deep in the hard surface and covered with glazed pebbles.[24] I counted eighty-four tracks running side by side. They had every appearance of being very old and must have represented a time when there were countless camel caravans in transit through this uninhabited region of today.

Exploring the ancient caravan route would have to wait for two and a half years. In 1955, Phillips returned with an armada of Dodge Power Wagons and followed the road a good twenty miles. He lost it in the sands that had drifted over it, then found it again. The Power Wagons bogged down. The caravan route led away into a no man's land of impassable dunes, six hundred feet high. With a melodramatic flourish, Phillips recounted: "From here on I knew we were through, for there is no barrier so great as billowing immeasurable sands stretching like a vast ocean as far as the eye could see in cruel and sublime grandeur."

But Phillips wasn't quite through, at least according to one of his bedouin guides. He apparently fixed on a particular red dune and proclaimed (for no apparent reason) that Ubar was under it. He shouted, "This is Ubar!" and, quick-drawing his pearl-handled revolver, emptied it in the air.[25]

When he returned to America, Phillips published his findings and prospered in the oil business.[26] But, though outwardly brash and cocksure, he had long been in frail health. He passed away at the untimely age of forty-two.

What a saga! The quest for Ubar had an *Arabian Nights* flair to it, a tumble of interwoven tales penned by scholars and scoundrels. And Ubar — if it existed at all — was still out there, undiscovered, a phantom city approached by a road that vanished in the dunes.

It was a city of dreams, or at least daydreams. Driving around Los Angeles, I would occasionally realize, with a start, that I no longer knew where I was; my mind, with increasing frequency, was lost in the sands. I would find myself puzzling over how to traverse the dunes. Maybe camels were the best after all. Or maybe specially designed vehicles. I ordered a catalogue from Johnnie's Speed and Chrome, an outfit that produced customized "sandrail" buggies that on a 45-degree slip face of dune could come to a full stop, then restart and climb on. The secret was the tires: huge, inflated with a minimum of air. They could run over you without leaving even a bruise. But then, I realized, these balloon tires would be quickly shredded by the Rub' al-Khali's intermittent flinty plains. And fine red sand would quickly clog the sandrail's exposed carburetors. Nevertheless . . .

We had decided to use the sandrails after all. And so far, so good. We had changed tires — from hard to soft, then back again — more than a dozen times, and now we were beyond where Wendell Phillips had given up and turned back. The dunes were enormous, but we raced up and over them with surprising ease. Then the wind picked up. We were in for a major sandstorm . . .

Where was I? Somewhere in Los Angeles, of course, but where? It was only when I looked in my rear-view mirror that I spied, two blocks back, the Denny's where I should have turned right.

Ubar . . . The sandstorm had passed, and we weren't at all sure where we were. We had strayed from the tracks of the Ubar road. Hoping to pick them up, we headed north and slightly west. We passed a small round boulder that seemed out of place in the dunes. We shifted into reverse and backed up. Turning the rock over, we found it scratched with ancient graffiti. Similar stones lay ahead, as did a great red dune. We scanned it with our binoculars and spied a fragment of masonry breaking free of its sands. Racing ahead, we discovered it to be part of a buried structure of finely cut stone. Carefully we removed a few blocks and entered a long, dark passage. It was clogged with sand, yet we could follow it deep into the dune. Our flashlights played across inscriptions. With A Dictionary of Old South Arabic (purchased at Hyman and Sons), we began to make sense of the elegant chiseled lettering . . .

No harm in dreaming.

3

Arabia Felix

He is crazed with the spell of far Arabia,
they have stolen his wits away.

The words of a poem by Walter de la Mare buzzed through my mind.[1] And it occurred to me: daydreaming of far Arabia aside, there was something very real I, as an amateur, could do to further the search for Ubar. Though prior seekers could not be faulted for their daring, it appeared that none had really done his homework. No one had taken the time — or perhaps had the opportunity — to see what, if anything, lay behind the campfire stories of the Rub' al-Khali bedouin.[2]

Given Kay's and my situation at the time (no contacts, modest means), this was about the *only* thing I could really contribute to the search for the lost city. The resources were certainly at hand: the nearby UCLA University Research Library alone had 60,000 volumes on the beliefs and lands of Islam. I could seek Ubar, not in the sands of far-off Arabia but in new and old accounts and documents. Was Ubar a real place? Or was it a mirage, a city that never was, a place that existed only in the realm of myth?

To begin with, was Ubar on any maps? For as long as I can remember, I've loved maps. As a kid, in my imagination I journeyed across them to distant isles and buried treasure. Now, presumably a grown-up, I planned to scour recent maps, then work my way back to the wonderful old ones that featured the woodcut legend "Arabia Felix."

To better understand the expeditions of the 1930s to the 1950s, I had already purchased a series of English operational navigation charts, the best available in the early 1980s. Designed for use by aircraft, they indicated prominent ruins with three little dots. I thought there was a remote chance that Ubar had been sighted and noted without anyone realizing what it was. But this was not the case. Though the rest of Oman was dotted with ruins (medieval or later), there was nothing whatever in the vicinity of Bertram Thomas's coordinates for the road to Ubar. The area was, in fact, blank. No contour lines, no shading. A legend said only "MAXIMUM ELEVATIONS BELIEVED NOT TO EXCEED 1800 FEET." Even in the early 1980s, the land was uncharted.[3]

That the landscape of the area had long been a blank was clear on maps going back as far as the 1500s. Huge swatches of desert were written off as "great Sandy Space" and "deserts très arides." The maps did note a number of old towns, survivors from antiquity, but not Ubar. An exception was the Reverend William Smith's 1872 *Atlas of Ancient Geography*, in which "Wabar" appeared in the middle of a surprisingly detailed map of Arabia. This was heartening, for it meant fabled Ubar was more than a recent bedouin invention.

Reaching further back, into medieval times, I couldn't believe my luck in finding a map that was all I could ask for. I first saw it as a reproduction, then obtained detailed slides of it from the British Library, where it resides. On the Psalter Map, a *mappa mundi* compiled circa 1225 and less than four inches across, tiny triangles marked the location of eighty-four of the world's major cities — among them, it would appear, Ubar! And what a city it must have been. Though it didn't appear on the Psalter Map by name, the area where the road to Ubar had been found in southern Arabia was marked:

are liberi ncoltime er culis

This says *are liberi n colime er culis,* Latin (not-very-good Latin, I was told) for "the altar of Liber and the Pillars of Hercules." In classical mythology, the Pillars of Hercules marked the edge of the known world, and "Liber" (often "Father Liber") was another name for Dionysus, god of the vine and wine, patron of revelry and ecstatic carrying-on.

But what were these two monuments, altar and pillars, doing in Arabia, let alone at Ubar? Dionysus was a Greek god, as was Hercules. And the Pillars of Hercules, I recalled, were said to have marked the Strait of Gibraltar. Delving into classical accounts, I pieced together what I thought was a plausible explanation.

First, consider Dionysus. According to the historian Diodorus Siculus, the god was born at Nysa, a "happy mountain" in Arabia Felix. It was only natural, then, that the Arabians should venerate Dionysus as one of their own, *especially* at a city known for its wanton ways. I found ancient Arabia's fascination with Dionysus confirmed by the Greek historian Herodotus: "The way they cut their hair — all round in a circle, with the temples shaved — is, they say, in imitation of Dionysus."[4]

Concerning the Pillars of Hercules, it seems that in the lore of the ancient world there were more than one pair. In particular, a chronicle of the conquests of Alexander the Great relates that Alexander found "Gates of Hercules" ninety-five days' march along the Babylon road, about what it would take a traveler to reach the Arabian Pillars of Hercules recorded on the Psalter Map. (Ever on the alert for spoils, Alexander ordered the pillars pierced to see if they were hollow or solid gold.)

It was late on a work night when I read this. Just for a moment or so, I closed my eyes.

Digging the great red dune was easier than we thought. Slowly but surely, we uncovered many buildings. Most had fallen to ruin, yet one was remarkably preserved. It was a temple. Two grand free-stand-

ing pillars dedicated to the god Hercules flanked its entrance. As described in an account of Alexander the Great's adventures, they were the equivalent of twelve cubits high.

Our hopes high, we passed between the pillars and entered the temple. It took several minutes for our eyes to adjust to the gloom inside. Quietly, hardly exchanging a word, we picked our way forward and were startled by the sight of a procession of drunken revelers reeling along behind the god Dionysus. They were figures on a frieze decorating a stone altar, figures frozen in time. We were awestruck. This had to be the very spot where, 2,300 years ago, Alexander the Great, the Macedonian king and conqueror, had come upon monuments to Hercules and Dionysus . . .

But alas, just when Alexander the Great became part of my theory concerning the Psalter Map, my theory fell apart. I discovered that though the Macedonian hero's conquests had been very real, they had given rise to some of the most outlandish fantasies of all time: the "Alexander books." Allegedly dating to an account written by one of his generals, these tales were popular well into medieval times. There were Armenian and Ethiopian Alexander books, an Indonesian version and an Icelandic one. They were forerunners of *Gulliver's Travels* and superhero comic books. In their pages Alexander encounters amazons, mermaids, and men who live on the smell of spices. He marvels at fleas the size of tortoises and lobsters as big as ships. He soars through the air in a griffin-powered flying machine and dives to the bottom of the Persian Gulf in a goatskin submarine.

It was likely, then, that whoever created the Psalter Map, probably a monk long on imagination (and short on spelling), had an Alexander book tucked under his straw pillow. And it turns out that the map's *are liberi n colime er culis* were not only fragments of spurious iconography, but they were inked in the wrong place. As the Alexander books have it, they should be in India; instead, they were set down in Arabia. There's a reason for this. In the Psalter Map, India is bi-

sected by a wall built by Alexander to keep the rapacious hordes of
the giants Gog and Magog from overrunning the world. The iconography of this takes up so much space that depictions of other
Alexandrian events, like his pillar and altar encounter, had to be
expeditiously shifted to the neighboring emptiness of Arabia, where
Alexander never set foot. In either reality or legend.[5]

The realization that the *are liberi n colime er culis* had nothing to
do with Ubar was naturally disappointing. It had taken several weeks
of spare time to decode the Psalter Map's promising inscription, find
it worthless, and then figure out why. I must admit, though, I enjoyed
the diversion. The Alexander books were surprisingly well plotted,
and wildly entertaining. For instance, in an Armenian version written
in the first person, Alexander, guided by the stars, crosses a desert that
is anything but deserted:

> The inhabitants of that place said that there are wild men and evil
> beasts there . . . There were men each twenty-four cubits tall; and
> they had long necks, and their hands and fingers were like saws . . .
>
> Moving on we came to a place where there were headless men.
> They had no heads at all, but had their eyes and their mouths on
> their chests, and they talked with their tongues like men . . . Then
> there appeared to us, about nine or ten o'clock, a man as hairy as a
> goat. I thought of capturing the man for he was ferociously and
> brazenly barking at us. And I ordered a woman to undress and go to
> him on the chance that he might be vanquished by lust. But he
> took the woman and went far away where, in fact, he ate her.[6]

In their bizarre way, the Alexander books were instructive, for
here was a good take on how myth worked. In the past I had read of
myth as "hieratic" or "teleodidactic," cryptic cultural constructs that I
could never quite grasp. Here myth was anything but arcane; it was a
lively, mischievous animal with scissor hands, barking at us, that
delighted in pouncing on the truth and making a merry hash of it. Yet

shards of truth survive. There *was* an Alexander, and he did have great adventures. The Alexander books were reasonably accurate in their depiction of *some* peoples, places, and events.

I understood, too, why myths persist across the centuries. They offer entertainment. They have an action-adventure quotient, they have an aura of wonder and mystery — and, best yet, they offer insights into the glories and fallibilities of the human heart, and how and why we live and die. As the conquest-obsessed, immortality-seeking Macedonian Alexander reaches the far side of his desert, two birds with human faces fly overhead and, in Greek, ask, "Why do you tread this earth looking for the home of the gods? For you are not able to set foot in the Blessed Island of the skies. Why do you struggle to rise to heaven, which is not within your power?"[7] These were sensible birds, not about to buy into Alexander's proclamation in 329 B.C. that he was a god. For all his might, chirped the duo, the Macedonian could not transcend his mortality.

Following the mythical footsteps of Alexander led me from the University Research Library to the gates of the Huntington Library in San Marino, near Pasadena, where I was kindly (and capriciously?) accepted as a "Reader," a researcher with formal privileges.

Set in magnificent grounds, the Huntington is a great marble building guarded by solemn Greek gods and housing a major research library of some 2 million books and 6 million manuscripts. Though famed for its holdings on British history and literature, the Huntington proved to have a surprising amount of material on Arabia: rare and wonderful editions of *The Arabian Nights*, the entire personal library of the great explorer-linguist-historian Sir Richard Burton, and, of special interest, a collection of original editions and manuscripts of the maps of Claudius Ptolemy.

In the late 1400s European printing houses sought to outdo each other bringing out woodcut editions of Ptolemy's atlas of the known world. Bologna, 1477 . . . Rome, 1478 . . . Ulm, 1482 . . . These *Cos-*

mographias, as they were called, were impressive. The editions at the Huntington were leather-bound and hand-colored, often in gold. Each turn of their vellum pages gave a whiff of the past, musty and mysterious. Locating "Tabla Sexta Asiae" — Ptolemy's map of Arabia — I saw that hundreds of sites and geographic features were accurately identified.

Yet understanding Ptolemy's *Cosmographias* was not a simple matter. It took me a while to grasp when and how they were compiled and what exactly they portrayed. To begin with, the *Cosmographias* were not at all what they first seemed. Though compiled in the 1400s, they were *not* the product of the Renaissance quest for knowledge and new horizons. Ptolemy was born in Greece and lived in Egypt circa 110–170 A.D. The Renaissance editions were, in substance, *reissues* of maps produced some thirteen hundred years earlier at the Bibliotheca Alexandrina, the Great Library of Alexandria.

Sometime around 150 A.D., as overseer of the Great Library, Claudius Ptolemy set out to map the known world. For information, he drew on his library's estimated 750,000 manuscripts, among them a number of "Peripluses" (literally, "round trips"), records of coastlines compiled by seafaring Greek traders. In the case of Arabia, these traders also brought back accounts of inland sites gathered, not firsthand but from local tribesmen. These informants measured camel journeys from place to place in "stages," each equaling a day's ride. Calculating that a stage averaged thirty to thirty-five miles, Ptolemy did his best to estimate the whereabouts of inland cities and towns.

To plot this and his other accumulated data, Ptolemy not only envisioned the world as round, but invented and set upon it lines of longitude and latitude. Every site was then given identifying coordinates. In Arabia, for instance, Medina (then called Yathrib, or Lathrippa) was at 71° × 23°, and Saba Regio, the royal city of Sheba, was at 73° × 16°.[8] In its original form, Ptolemy's atlas — including his map of Arabia — was a wonder of the world. Nothing so complete, so detailed, so accurate, had been done before. And therefore it was a very sad day when, in 391 A.D., at the order of the Roman emperor Theodosius I, a religious mob trashed and torched Alexandria's library — and Ptolemy's atlas went up in smoke.

Nonetheless, fragments of Ptolemy's work survived. At least one copy of his table of coordinates — his listings of landmarks and sites — was saved and passed down through the centuries, until in the late 1400s European mapmakers laid out longitude and latitude grids and plotted afresh Ptolemy's coordinates of coastlines, mountains, rivers, and tribal fiefdoms. They marked his cities and towns with quaint castles or little dots, often in gold. They reconstructed, quite successfully, the world as Ptolemy knew it, as it was not long after the time of Christ.

On Ptolemy's map of Arabia — if anywhere — I should find Ubar. And sure enough, on most editions, the tribal name "Iobaritae" — Latin for "Ubarites" — appears more or less where Bertram Thomas encountered his road to Ubar. But there was no identifiable *settlement*, only evidence that an Ubarite tribe once may have wandered the region's sandy wastes. No castle or golden dot on Ptolemy's map, no city. And it wasn't as if I could look further into the past. Before Ptolemy, the only maps were very crude and usually local.

For several weeks there seemed no way to get beyond this impasse. Then, to better understand how Ptolemaic maps were constructed — and to attempt to conjure something out of nothing — I decided to make one of my own, step by step. Working from a table of coordinates printed in Ulm, Germany, in 1482, I plotted nearly four hundred landmarks and towns, just as Renaissance mapmakers had done. The project, which took several evenings, was intriguing, much like working a jigsaw puzzle. But there were no surprises. Except . . . a day or two after I'd finished, one of the places I'd plotted popped up in my mind and bothered me. Omanum Emporium, "the marketplace of Oman," appeared in *western* Arabia at 77° × 19°. As far as I knew, the

key longitude coordinates

70   78 80   87   90

Lathrippa Arabia

Omanum
Emporium

IOBARITAE

Felix

Sabe Regio

Ptolemy's map of Arabia (simplified)

ancient land of Oman was in *eastern* Arabia (as today's Sultanate of Oman is).

What, then, was "the marketplace of Oman" doing so far from home? The next time I could make it over to the Huntington, I doggedly double-checked the library's many editions and manuscripts of Ptolemy's *Cosmographia*. Nothing to be learned. When I returned the atlases to the rare book desk, I felt a little guiltier than usual for asking the librarians to lug these enormous volumes back and forth from the Huntington's vaults. It was a little before five, closing time.

I hesitated, then drifted back to the library's main reading room. Though the lights were off, its oak paneling glowed in the last rays of the winter sun. I noticed, tucked under a shelf, a version of Ptolemy's atlas that I had thumbed through but not really studied. It was not an original edition but rather a reproduction, printed in the 1930s, of the Ebner Manuscript owned by the New York Public Library.

The Ebner Manuscript was a one-of-a-kind work, hand-done in 1460. It predated, I realized, all the other atlases at the Huntington. I turned to its map of Arabia. Omanum Emporium was in its usual place, at roughly 78° × 20°. Then I flipped to the atlas's table of coordinates and found Omanum Emporium listed at 87°40' × 19°75'. Something was wrong. *Omanum Emporium's longitude was listed as 87° in the atlas's table of coordinates, yet placed at 78° on its map of Arabia.* A door swung open. A library guard looked in, glanced at his watch, and cocked his head at me, the only person left in the cavernous room.

"Just a second, just wrapping it up," I said . . . and checked the Ebner Manuscript's map of Arabia to see where the 87°40' × 19°75' listed in the table of coordinates might be. It fell in eastern Arabia, *just above the word "Iobaritae."* Here was a settlement — long misplaced — in the homeland of the Ubarites! And it was a major settlement, for in some editions of Ptolemy's map of Arabia, a dozen or so cities are singled out (sometimes by larger type, sometimes by an asterisk) as being particularly important. Omanum Emporium consistently makes the cut.

What exactly happened in the monastery of Ebner in the 1460s is not clear. But a likely scenario is this: the creator of the manuscript correctly copied an earlier list of Ptolemy's coordinates and, working from these coordinates, set to drawing his map of Arabia. He outlined the coast, added mountains and rivers. One by one, he filled in the cities and towns. When he got to Omanum Emporium he ran out of ink . . . or it was time to light a candle . . . or perhaps he paused to brush away the scriptorium cat. One way or another, the scribe

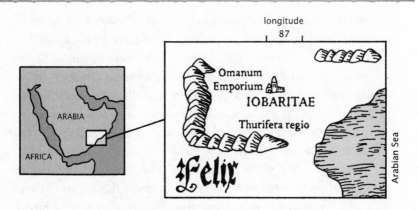

Detail of Ptolemy's map of Arabia (corrected)

slipped up. When he put pen to parchment, the 87° latitude of the table of coordinates became 78° on the map. This simple inversion placed Omanum Emporium far west of where it belonged, a mistake that appears to have been repeated on later maps copied in Bologna, Rome, and Ulm.

Heading home that afternoon, I was barely aware of what freeway I was driving. Was "Omanum Emporium" Ubar itself? In the ancient world, sites frequently had several names, just as they have in modern history. What was once Niew Amsterdam is Manhattan, Gotham, the Big Apple. Yet Omanum Emporium did not rest easily at its new location. What was a "market town of Oman" doing way out in the desert? An emporium was typically a seaside trading town.

Back at the Huntington, I found the answer no more than a half inch away on Ptolemy's map. Close by Omanum Emporium's newly established site was the legend *Thurifero Regio*, or Incense Land. In the ancient world incense was a major commodity, in demand for both temple and household use. Frankincense, I had read, was the resin of a humble desert tree, which, by the time it reached the faraway markets of the Mediterranean, was as valued as gold.

Omanum Emporium — Ubar — could well have had a role in

the harvest and trade of incense. In fact, understanding how the trade worked might clarify the city's reason for being, and its rise and fall.

Iobaritae . . . Omanum Emporium . . . Thurifero Regio . . . These three legends on Ptolemy's map appeared to equal a tribe, a city, a trade in incense. And it intrigued me that in all of Arabia, Omanum Emporium was *the sole major site on Ptolemy's map that has not been accounted for.* So it would make a great deal of sense if it and lost Ubar were one and the same.

Before the gates to the Huntington swung shut, I had time for a walk in my favorite of the library's gardens, the cactus garden. Though it was winter, pencil and silver cholla cactus were in bright yellow and red bloom. Admiring them, I reflected on my journey across the deserts of many maps of Arabia, new and old. With no little difficulty, I had made it past the saw-fingered man and the barking goat-man and had stood in awe of the landmark achievement of Claudius Ptolemy, the world's first great cartographer.

I looked back at the Huntington and realized why scholars, for one very good reason, had never taken Ubar seriously: the city wasn't on Ptolemy's map. Yet — under another name and in the wrong place — it had been there all along.

4

The Flight of the *Challenger*

FOR THE TIME BEING, at least, Ubar's place and purpose made sense. In a far corner of ancient Arabia, incense was harvested, then taken via a time-worn road to the fabled caravansary of Ubar, known to Claudius Ptolemy as the marketplace of Oman. The incense would then have been loaded on camels for a daring journey across the Rub' al-Khali to the great markets of Petra, Alexandria, Jerusalem, Damascus, Rome.

But now, could the search for Ubar be defined, narrowed? Ptolemy's map of Arabia was fine for determining the lay of the land, yet its inherent distortions made it impossible to zero in on a site with any degree of accuracy.[1] The best that could be said for Ubar was that it was somewhere in a 100-by-300-mile sandbox, somewhere in 30,000 square miles of gravel plains and dunes.

A while back, the *Los Angeles Times* had had an article about an airborne radar system that had successfully located Mayan ruins hidden beneath a dense jungle canopy. The article said that a similar radar system, in the near future, was to fly aboard the space shuttle.

It was an improbable idea, but maybe the shuttle could look for Ubar. It couldn't hurt to ask. On a Thursday in 1983 I took a deep breath, called Pasadena information, then dialed the main number for NASA's Jet Propulsion Laboratory.

"Good morning. JPL-NASA. May I help you?"

"I'd . . . I'd like to talk to someone about using the space shuttle to look for a lost city," I asked in an unconvincing this-is-just-a-routine-inquiry tone of voice.

"Oh . . ."

A pause ensued. "A lost city in Arabia," I blurted, as if this validated my question.

"I'll connect you with Dr. Blom," the operator said in a rush, and was off the line.

An extension rang. Who was this Dr. Blom? The duty officer for fielding dreamers of Atlantis, hollow-earthers, the flying-saucer crowd?

"Ron Blom," said an affable voice.

Introducing myself to Dr. Blom, I managed to slip in that I had worked for the National Geographic Society and Walt Disney, hoping this would give me some shred of credibility. I think I scored a point or two as I (correctly!) alluded to the orbital parameters of the shuttle's flight path. "Could we talk further?"

"I don't see why not," replied Dr. Blom. In fact he was free for lunch that very day, provided I made it to JPL by 11:30. Scientists like to eat early.

Not far from the Huntington Library, the Jet Propulsion Lab also has serene, grassy grounds and a population of serious researchers, at the time more than ten thousand people working on NASA projects. As scholars at the Huntington probed secrets of the past, JPLers looked to the future, to revelations brought by the manned and unmanned exploration of space.

Ron Blom was clearly a serious scientist. A Caltech-trained geologist, he had a beard and favored plaid shirts and sleeveless sweaters. Yet there was an air of bemusement about him. At the JPL cafeteria, he thought I might enjoy watching rocket scientists work the salad bar. For $2.25 they could help themselves to whatever would fit in a little styrofoam bowl. They began by laying solid foundations of garbanzo beans, baby tomatoes, beets, and the like. Upon this they built edifices of romaine and spinach. Buttressed by celery and carrot sticks, their leafy constructs soared higher and yet higher, triumphs of structural engineering. As I headed off to the checkout counter, two scientists were debating the viscosity and slip-face factors of green goddess versus California ranch dressings.

"On me, Dr. Blom," I offered. This was a nice gesture, I thought. And it would have been, had not my wallet been empty. In my haste to get to JPL, I had forgotten to stop at a money machine. Not only was I going to take up Dr. Blom's time with a harebrained idea, but he would have to pick up the tab. I tried to make the best of it.

"Thanks, really. Dumb of me . . . But how about this: the day we land in Oman to look for Ubar, I'll return the favor. I'll buy you lunch."

I was able to make good on this — seven years later. In the meantime, Ron was to prove a stalwart of what became an Ubar team. We spent a good part of that first afternoon chatting in his office, where I was put at ease by the message on a Post-It stuck above his computer terminal: "DARE TO BE STUPID!"

We started with my Operational Navigational Chart J-7, on which I had marked what Bertram Thomas and Claudius Ptolemy had noted about the possible location of Ubar. Ron was amazed that in our day and age this area was still left white, unexplored and uncharted. He filled me in on the system that could remedy the situation and perhaps even find Ubar.

Scheduled to fly on a shuttle mission eighteen months thence, SIR-B — Shuttle Imaging Radar B — would map selected areas of the world by beaming down a powerful microwave and then digitally recording its bounced-back return signal. The radar had a unique ability to see through cloud cover and dense vegetation. It could even penetrate dry sand and reveal buried features, both natural and man-made. It could create images of sights unseen, of times past.

An earlier version of this radar, SIR-A, had been borne aloft on a

1981 mission of the shuttle *Columbia*. Ron had been a principal scientist on the project and savored relating the story of some of the first images to be analyzed. Taken over Africa, they were of northern Sudan's pancake-flat Selima sand sheet.

"The immediate thought was that somebody had grabbed the wrong roll, that this was really someplace else," Ron recalled. "The radar images didn't look anything like what we understood to be the Selima's bland surface."

It was the right roll. What those images recorded was not the veneer of the windblown sand sheet but a denser surface buried beneath it. The radar saw right through the sand as if it weren't there, revealing a hidden landscape of streams and rivers, now dry and buried but once filled with the torrential runoff of monsoon rains. The radar had reached back hundreds of thousands of years and taken a snapshot of a vanished land.

Ron's team traveled to Egypt and Sudan to "ground-truth" the Selima images. In a totally featureless landscape, they plotted where on their radar image two buried rivers converged — a good place for a campsite. They dug down through the sand and, sure enough, found riverbank contours and Stone Age artifacts.

"With Ubar, say it was buried. We could pick up what?"

"A lot, as long as it was reasonably near the surface."

"Meaning?"

"Six feet down, no problem. Theoretically, the radar can penetrate up to five meters — eighteen feet, that is — of very dry sand."

"And what would we see? Walls? Buildings?"

"It would have to be something pretty sizable. But sure, structures should show up, as long as they were on the order of thirty meters across. That's the limit of SIR-B's resolution . . . the size of a pixel."

I nodded, not entirely sure what a pixel was.

Walking me back to the lab's main gate, Ron promised to arrange a meeting with Charles Elachi, who not only headed up JPL's Radar

Imaging Geology Group, but was director of the entire ambitious NASA SIR-B shuttle project.

A week later we were in Charles's office. And if there was an occasion for intimidation, this was it. Here was a scientist's scientist, at thirty-six a holder of Ph.D.s in planetary geology, electrical engineering, and quantum physics. He had also grown up in Lebanon and was well aware that in the Middle East fact and fancy often became hopelessly intertwined. I admitted this to be the case with Ubar, but said I felt that, as sources, Bertram Thomas and Claudius Ptolemy could be trusted.

"Okay, how about you?" Charles said to me. "You think this place Ubar really exists?"

I had to be honest. "Frankly? I don't know."

"That's a perfectly good answer," Charles replied, without a hint of hesitation. "What's science for, if not to find out what exists or doesn't. Right, Ron?" Ron nodded, and Charles continued, "But say we did this, we went looking for Ubar with our spaceship, it would have to be unofficial, you understand. With you, we're not exactly dealing with . . ."

I could complete the sentence: ". . . an academic institution."

"No offense, you understand, It's just . . ." Ron interjected.

"Oh, my, no offense at all," I assured them. "Listen, I'm amazed you're even considering this."

"What if," Ron suggested to Charles, "we made Ubar a target of opportunity? If we're not in trouble on anything else, that is."

"We could. Yes, we could certainly do that," Charles agreed, and the meeting was over.

As Ron and I walked away down one of JPL's endless corridors, I wasn't quite sure what the upshot of the meeting was, but I suspected that the space shuttle's trajectory might just happen to intersect with the road to Ubar.

"Was what happened in there what I think happened?" I asked.

Ron smiled. "Uh-huh."

A few weeks later, a manila envelope arrived in the mail from JPL. In it was a computer-generated map of Arabia. Two parallel lines, fifty kilometers apart, angled across the peninsula . . . the paths of two scheduled space shuttle passes over the Ubar area. A Post-It from Ron said simply, "I think you might find this of interest."

Several months later, I intently followed the *Challenger's* radar mission.

Friday, October 5, 1984 . . . At 7:03 A.M., a blaze of light pierces the dark Florida sky. A huge cloud of smoke billows out across Cape Kennedy's swampland. In a rare night launch, the space shuttle *Challenger* thunders skyward, punches a hole in a cloud, and accelerates into orbit. It is a perfect launch, the sixth for the *Challenger*. A commentator notes that it is "now a fully mature spaceplane."

Saturday . . . The shuttle mission isn't going so well. One problem is an out-of-control, wildly swinging KU-band antenna, designed to relay data from the SIR-B radar to a Tracking and Data Relay Satellite (TDRS), which is supposed to beam the information back to earth. In Houston the press is told that SIR-B may be able to cover only 20 percent of its targets. There would have to be drastic prioritizations. Goodbye, Ubar. The flight director, John Cox, says that "Murphy has a way of getting to you," Murphy, of course, being the author of the law "Whatever can go wrong, will."

Sunday . . . Hopes rise when Mission Specialist Kathryn Sullivan becomes the first American woman to walk in space. Working with Specialist David Leetsma, she stops the KU antenna from reeling about like a drunk at a party. They lock it in a fixed position. Priority SIR-B data can now be downloaded by using maneuvering jets to point the *Challenger* at the TDRS satellite, a space-age version of the tail wagging the dog. As Charles Elachi explains, "It's like rotating your house to re-aim the TV antenna and get better reception."[2]

Though a controller announces, "We're back in business," the

SIR-B array malfunctions again. As part of "the revolt of the anten-
nas," the radar's main thirty-five-foot antenna won't budge from the
cargo bay. Finding this unacceptable, Sally Ride pokes it with a
mechanical arm, and it springs to life. Next a short circuit plagues
and intermittently degrades SIR-B's images. There's no way to fix it.

Tuesday . . . It is touch and go with the SIR-B array.

Wednesday . . . Still touch and go.

Thursday . . . On its 96th orbit, the *Challenger* streaks over the
Ubar search area, presumably without acquiring any data.

Friday . . . The spacecraft's 112th orbit again takes it over our search
area, on a path fifty kilometers west of the pass the day before. Six
hours later, dodging Hurricane Josephine, Captain Robert Crippen
safely guides the *Challenger* back to earth.

In a NASA news conference, it is estimated that the SIR-B radar
achieved only 40 percent of its goals. Charles Elachi allows he is "a
little bit disheartened," but is confident SIR will fly again another
day. The next week's issue of *Time* reports: "The loss of viewing time,
coupled with the antenna problems, meant that a few scientific projects had to be sacrificed, among them a hoped-for image of the
Arabian Peninsula near Oman, thought to be the site of an ancient
lost city. Shrugged Charles Elachi, 'The lost city will have to be lost
for another year or so.'"[3]

Ron Blom, Kay, and I were as disappointed as Charles was. About
the best we could hope for, he said, was a radar reflight in three or
four years' time (actually, it would be a full ten years before SIR-C
went aloft).

Ron went on with his work interpreting the geology of America's
western deserts from space, Kay continued to combat crime as a federal probation officer, and I made documentaries for television. Still,
whenever I could spare a few hours, I wistfully returned to either
UCLA's University Research Library or the Huntington. I pursued

the possibility that the city was a link in Arabia's incense trade, perhaps even its point of origin. But I had relatively little idea who might have lived there or what had become of them.

Now I had a new lead. As I chased after Ubar, I had frequently bumped into references to another lost city in Arabia, spelled either "Irem" or "Iram." The Koran asks, "Have you not heard how Allah dealt with 'Ad? The people of the many-columned city of Iram, whose like has not been built in the whole land?"[4]

The legends of Ubar and Iram, I learned, had much in common. In fact, they had too much in common. Both were allegedly feckless cities destroyed by an angry Allah. They flourished and fell at the same time. Moreover, they were in the same area, a region of the Rub' al-Khali known as the desert of al-Ahqaf.[5] Most telling, *they were both built by the same tribe, the shadowy "People of 'Ad."* Was Iram another name for Ubar? Yes, I thought.

This identification had a bonus: far more was written about Iram than about Ubar. An early faint remembrance of a place and its people may be embedded in the meaning of the proper names "Iram" and "'Ad."[6] At its Semitic root, "Iram" means a pile of stones erected as a way marker, and "'Ad" is most likely the same as "Gad," the god of fortune once revered throughout the Semitic world. Thus we might conjure up an image of a city that was a landmark in an unknown land, and an image of a people pursuing fortune, a people enamored of wealth and its trappings.

This, of course, was a major conjectural leap, yet when Iram and 'Ad appear in surviving pre-Islamic poetry — a mirror of life in ancient Arabia — there is more than a hint that these people were living a worldly life and having a good time of it. One poet pondered what it would have been like "had I been a man of the race of 'Ad and of Iram":

> Roast flesh, the glow of fiery wine,
> > to speed on camel fleet and sure . . .

White women statue-like that trail
   rich robes of price with golden hem,
Wealth, easy lot, not dread of ill . . .[7]

Fragments of other poems imply that the People of 'Ad — and even their camels — were a wicked lot. Seeking a superlative for nastiness, one poet comes up with: "of ill omen, eviler than Ahmar of 'Ad." Describing the aftermath of warfare, another intones: "She [War] brought forth Distress and Ruin, monsters full-grown, each of them as deformed as the dun camel of 'Ad."[8]

It is interesting that in pre-Islamic poetry, which dates back to 500 A.D., perhaps earlier, the 'Ad appear as a past-tense people, gone from Arabia and even from the face of the earth. What became of them? And of their city? The prophet Muhammad knew. Preaching in Mecca around 640–650, he declaimed: "Arrogant and unjust were the men of 'Ad. 'Who is mightier than we?' they used to say. Could they not see that Allah, who had created them, was mightier than they? Yet they denied our revelations. So over a few ill-omened days, We let loose on them a howling gale, that they might taste a dire punishment in this life; but more terrible will be the punishment of the life to come."[9]

Muhammad's preachings, gathered into the Koran, make much of Iram's destruction by a "hurricane bringing a woeful scourge," a sudden and dramatic end for the People of 'Ad: "When morning came there was nothing to be seen besides their ruined dwellings. Thus We reward the wrongdoers."

In the Koran's account of Iram's violent demise, two major figures appear: the worldly King Shaddad and the prophet Hud, who warns the 'Ad that their wicked ways will bring them to a bad end. Regrettably, little more is said of either character or of the full story of Iram. In the Koran, tales are never told from beginning to end as historical accounts; rather they are repeatedly — and fragmentarily — cited to drive home a moral point. For the 'Ad, the moral point is a thunderous warning: "You squandered away your precious gifts in your earthly life and took your fill of pleasure. An ignominious punishment shall be yours this day, because you behaved with pride and injustice of the earth and committed evil . . . Serve none but Allah. Beware of the torment of a fateful day."

In the Koran, Muhammad gave the People of 'Ad a resounding stamp of disapproval — and thereupon licensed a host of Islamic historians, geographers, travelers, and storytellers to decry the city and people of Iram. What better theme than comeuppance for the wicked? By medieval times, the tale of the city's rise and fall had been told and retold dozens of times and had been seized upon and woven into the fabric of the *Arabian Nights*. Along the way there was considerable embroidery. Incidents, details, and characters were added, some imaginary to be sure, but some — maybe — harking back to forgotten documents or preserved oral traditions.

As a late-night project, I compiled a cast of characters and genealogy for Ubar. Whereas the Koran mentioned only King Shaddad and the prophet Hud, my tale came to have more than thirty interrelated players. There were forebears of King Shaddad dating back to Noah. There was the sage Luqman ibn 'Ad, a pair of dancing girls known as "the Two Locusts," and the woman Mahdad, the first victim of the havoc of the city's final days. My genealogy even had a column for "nisnases," weird, monkeylike creatures said to have taken over and haunted the ruins of Iram/Ubar after the place was divinely demolished. They dated back to a common ancestor, one al-Nisnas ibn (son of) Omain ibn Aalik ibn Yelmah ibn Lawez ibn Sam.

With no other way to seek Ubar, here I was plotting a family tree of creatures described as having one eye, one arm, and one leg!

Myth's path, I knew, was uncertain and abounding in slippery slopes. But give the myth of Iram/Ubar credit for staying power. Though the tale achieved its peak of popularity in the 1100s and 1200s A.D., it was still very much a part of Arabian culture when outsiders penetrated the peninsula. In the early 1800s Johann Burckhardt wrote of the region of al-Ahqaf: "According to the tradition of the Arabs, this desolate region was once a terrestrial paradise, where dwelt a race of giants, who, for their impiety, were swallowed up by a deluge of sand." In 1860 Colonel L. Du Couret, a wanderer through Arabia, recorded the venerable story of how, deaf to the prophet Hud's message, "the 'Adites continued to abandon themselves to the practice of an idolatry the most besotted."[10]

Could a story so long-lived, so rich and complex, be based on nothing? Yes, it could, given its cheerful embrace of nisnases and the like. But, myth, it's widely believed, nearly always springs from something — a place, a person, an event — that once was as real as a hometown, an ancient king or prophet, or a wrenching disaster.

In the late spring of 1985, I called Ron Blom at JPL. Even though the search from space was on indefinite hold, I felt he would be interested in the Ubar-Iram link. It had produced new clues: evidence of the city's existence dating to the dawn of Arabic literature, a location in the al-Ahqaf region, and the possibility that the city had come to a violent end between the time of Ptolemy's map and its mention in pre-Islamic poetry, between 150 and 350 A.D.

"Funny you should call," Ron said, "because I've got something for you. You might want to stop by."

Later that week I did. The "DARE TO BE STUPID!" Post-It still stood watch over his computer. Certainly done that, I thought, as I showed Ron my schematic, rather outlandish Ubar genealogy. In exchange, he picked up a manila envelope. "Down the hall," he said, gesturing, and led the way to a dark, windowless room. He switched on a large light box, then from the envelope matter-of-factly produced a black-and-white transparency.

"Can't see much on this one. Transmission efficiency was way down, only about two percent. This line over here looks to be a pipeline . . ."

He placed a second transparency on the light box, just below the first. "Better resolution on this one. See the dunes here?"

I also saw that the transparency had a printed notation: "JPL DATA TAKE 96.1." We were looking at the results of the space shuttle *Challenger*'s flight over Arabia! Ron smiled and explained that really these weren't supposed to exist. But somehow, in the chaos of the ill-starred radar mission, they had been downloaded to the TDRS satellite, from there to Maryland's Goddard Space Center, and finally to JPL.

"How about this? The radar's behaving now."

The image was of the edge of the Rub' al-Khali, where its sea of dunes gave way to gravel plain. It clearly mapped the heart of our Ubar search area. Further, the radar had seen through superficial drifting sand, even small dunes, and revealed a landscape thousands of years old. It was a landscape where rivers once flowed, where lakes once formed.[11] A landscape of Arabia as a vast savanna, an Eden even. Certainly not a desert. Ron pointed out a particularly large lakebed, over twenty kilometers long, the size of California's Salton Sea. On its shores early man could have camped, hunted, even fished. Here civilization could have gotten a toehold and held on, even as the surrounding terrain became desert and the lake vanished.

We calculated the point where Bertram Thomas had crossed and regretfully left behind the road to Ubar. It was in the middle of Ron's long lake. With a magnifying glass we searched for the road, but found no trace of it. As in the line from *Winnie the Pooh*, "The more they looked, the more it wasn't there."

"Right . . . ," Ron said with a sigh. "To be expected, I guess. With space imaging, it's rare you find what you're actually looking for." He added, "But then you come upon something else that may — or may not — be helpful."

As he looked at the shuttle's transparencies this way and that, Ron spoke of what might be described as the Zen of Space Imaging. You

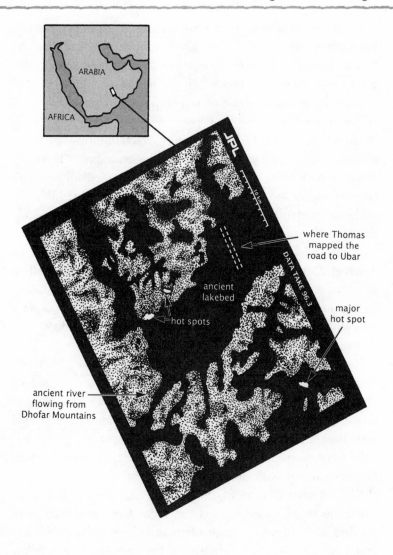

Radar image of area of the Ubar road

clear your mind of preconceptions. You look for an anomaly, something out of place. If you're interested in the presence of man, you look for geometric features. Straight lines, right angles, and the like. But beware: nature can concoct shapes that you could swear were roads, walls, or canals. Above all, you keep an open mind.

Ron pointed out a few "hot spots," white patches where the radar may have recorded "disturbed earth." Such disturbances, he explained, were often caused by man and his endeavors. "These are places worth looking at. If we ever could." He singled out a particularly striking hot spot. "Like here. It's a hill. But why so bright?"

He gathered up the images and handed them to me. "Have a look. See what you can find."

At home I spent hours poring over the images, millimeter by millimeter, until at last, south of the large lakebed, I saw the outline of . . . a city? With a macro lens I photographed this suspicious feature and blew it up so that I could examine its every dot of grain (its every digital pixel). Sure enough, I saw a distinct continuous line — a wall? — enclosing a mile-square area.

Hopes up, I returned to JPL, only to have Ron say, "Artifact." Radar imagery, he explained, has the perverse ability to produce random geometric features — called artifacts — that have nothing to do with what is really there. But I shouldn't feel bad; artifacts had led more than one serious scientist down the garden path.

In archaeology, artifacts are keys to the past. In space imagery, they're meaningless doodles.

We met with Charles Elachi and reviewed the SIR-B images. There was no sign of the road to Ubar. Yet we had a good overview of the area and at least one particularly promising hot spot. We had found just enough to want to find more. Ron suggested that we follow up on the SIR-B sweep of Arabia by obtaining satellite coverage. It would be in color and could be processed to bring out features not visible on the SIR-B radar images. Charles agreed. But this would take time, he cautioned. JPL was swamped with NASA priority projects.

As the meeting broke up, I told Ron and Charles of Abu Muhammad al-Hamdani, a southern Arabian scholar who had spent a lifetime (893–945) gathering reports of past civilizations. In his *Eighth Book of al-Iklil,* he lists "Iram, the city of Shaddad ibn 'Ad" as first among the lost treasuries of Arabia; he predicts that someday "it will be unearthed by ants. . . . This will take place when despots are gone and the tyrannous pharaohs are no more."[12]

"Well, there are no more pharaohs," Charles reflected. "But despots? In that part of the world it could be a while."

I decided not to tell Charles and Ron of an additional Ubar item, a curse recorded in *The World History of Rashid al-Din* (1290): "Whoever shall find and enter Ubar will be driven mad with fear."[13]

The Search Continues

IN THE EARLY MORNING of January 28, 1986, seventy-three seconds after launch, the space shuttle *Challenger* came to a sudden and fiery end, killing six astronauts and Christa McAuliffe, a high school teacher from New Hampshire who was to give school lessons from space. After a period of shock, disbelief, and sadness, Kay and I and our JPL friends rarely discussed the accident, yet we tacitly agreed that we owed the ship and its crew our best effort in the search for Ubar. One way or another, we would travel to Oman and walk the desert mapped by the *Challenger*'s 1984 radar images.

Kay and I agreed that the time had come to see what could be done to organize a ground expedition. It was also time to get some sound advice from professional archaeologists. And so, a month after the *Challenger* went down, I flew to Montreal for a rendezvous with Sir Ranulph Fiennes, Ran for short. He was en route to the high Arctic for another installment of his polar adventures.

Though a hereditary baronet, Ran Fiennes's life had been anything but tea and crumpets. In the military he had shown promise as an officer in the Special Air Service — until the night when he and his mates got themselves up in greasepaint and camouflage and dynamited the principal set of the movie *Dr. Doolittle*. It wasn't right, he felt, for the film's producers to push around the people of Castle Combe, a picturesque village near his SAS barracks.

Parting ways with the military, Ran took up a life of professional adventure. He ran the upper Nile in a hovercraft, rafted British

Columbia's treacherous Headless Valley, and from 1978 to 1981 led an epic expedition to circumnavigate the globe on the Greenwich meridian, crossing both the South and North poles. As a London cabby put it to Kay and me, "Ran Fiennes . . . that boy will do anything to avoid an honest day's work."

I had worked on *To the Ends of the Earth*, a documentary on the Greenwich meridian expedition, and I knew Ran as an immensely likable, if sometimes quicksilver, fellow. I also knew that, although best known for his polar exploration, he had, in his military days, seen service in Oman as the leader of an irregular bedouin patrol. He knew the country and its people. He spoke Arabic. He had heard of Ubar. In his book *Where Soldiers Fear to Tread*, he wrote:

> The bedu tell of such places around their camp fires but none can point accurately to the ancient sites. Their ancestors passed on tales of sand "yetis" that moved with great speed and grace but were hideous to behold, having only a single leg and arm attached to their chest. Their home was the epicentre of the Sands, that mysterious place where no bedu had ever been and where the lost city of Ubar was to be found.[1]

Ran was intrigued by what I had learned of Ubar, as well as the potential of further JPL space imaging. We discussed leading an expedition together. I would continue doing research and work with JPL on a plan for locating the elusive city. For his part, he would seek the blessing of Oman's Sultan Qaboos ibn Said, whom he knew, and would then plan and oversee the expedition's logistics. But there was a hitch. He would not push ahead until I had come up with the necessary funds. He guessed that our expedition would cost $35,000, maybe more.

With a wave of his ice axe, Ran was off to polar reaches. With some trepidation, I continued on a round of meetings with East Coast experts on Arabia. Trepidation because what I proposed doing — searching for a site by relying on historical clues — was anything but archaeologically p.c. This had a lot to do with a number of past scholars who, guided by the Bible, had for over a century wandered the Middle East seeking the actual sites of biblical revelations, battles, and the like. In spite of all the money spent and the hopes of the faithful raised, their approach had not been terribly productive. In fact, it had produced such a muddle of speculation and misidentification that today Middle Eastern archaeologists tend to ignore (or at least mistrust) historical references and clues and focus on what they find in methodical, dispassionate surveys.

Perhaps the archaeologists I met still secretly liked the old romantic, if not very effective, way of seeking a site. Or perhaps they were being tolerant of an amateur. In any case, they enthusiastically supported the idea of an expedition to find Ubar. At Brown University I met with professors Ernest Frerichs and Jacob Neusner. They told me that historians had unjustly ignored Arabia and that Ubar — if it existed — might have a significant role in the Middle East's complex archaeology. Gordon Newby of the University of North Carolina, an expert on early Arabic texts, was quite familiar with Iram/Ubar. He was particularly intrigued by the prophet Hud, a name that linguistically could be taken to mean "He of the Jews." He wondered: could Hud have been a lone wandering Israelite, a voice in the wilderness decrying Arabia's idolatry?

In Washington I spent an encouraging afternoon with Smithsonian archaeologist Gus Van Beek. I also visited with explorer Wendell Phillips's sister, Merilyn Phillips Hodgson, who, since his death, had supported Arabian archaeology. She arranged for me to meet Father Albert Jamme, the inscription expert who in 1953 had prompted a bedouin sheik's unreasonable, insatiable lust for latex, which in turn had propelled the Phillips expedition on to Oman and the search for Ubar. She said not to take it personally if the learned Jesuit threw me out.

Tucked away under the eaves of a timeworn Victorian mansion on the campus of Washington's Catholic University, Father Jamme's office was crammed with arcane journals, latex squeezes, and worn oaken files indexed not in English or even his native French but in ancient south Arabian script. "A to Ag," for instance, was ⴼ-ⵏⴼ. I knew that Father Jamme had no patience for fools and that many a visiting scholar had been sent packing down his narrow stairs with fulminations of "dangerous assumptions" and "unbelievable ignorance" echoing in his ears. It was a relief, then, after a few uncertain minutes, to be unrolling the father's annotated maps of Arabia and relating them to JPL's space images, which he found of great interest.

For some curious reason, members of his order — beginning in the 1860s with the self-described Jewish Jesuit, Gifford Palgrave — had long had a role in penetrating the mysteries of Arabia. And to Father Jamme, the Ubar region was a critical missing piece in the puzzle of an ancient land.

"The road!" he exclaimed as he paced about. "The road to Ubar! Yes, it could well be! An expedition? Yes! It will be valuable, even if it's to show us there's nothing there!" (This might be his idea of a crackerjack expedition; it wasn't mine.)

Father Jamme carried on about the importance of tracking the trade routes of ancient Arabia, pointing out that the land beyond Dhofar's coastal mountains was still pretty much an archaeological blank. He told me that according to classical sources, the world's finest incense — a translucent "silver" grade of frankincense — had been harvested on the back slope of those mountains, taken down to the coast, and exported by sea.[2] But if evidence could be found that frankincense was also transported directly across the desert via Ubar, a new — and until now secret — chapter of the ancient past might be revealed.

After I returned to Los Angeles, we corresponded. In one letter Father Jamme said that although no known inscriptions mentioned the place, he hoped that someday I might come across the simple word)ⵏⴼ, Ubar. Along the way, though, I had better not make any "dangerous assumptions."

On a practical level, Kay and I wondered what we could do about Ran Fiennes's challenge: raise the money, and he would come aboard. I knew that whatever my abilities, fundraising wasn't one of them. I thought of George Hedges, an attorney I had worked with on an interactive video project. George, a complex and talented fellow, could not only raise money, he *liked* raising money. There were many Georges: George the dramatic trial lawyer, who often worked pro bono on near hopeless death penalty appeals; George the ex-rocker (who once had been Lush Pile of Lush Pile and the Car Pets); and George the archaeology buff, who had excavated sites in Greece and held a master's degree in classical languages from the University of Pennsylvania.

George and I agreed to work together to try to get a modest Ubar expedition off the ground.

In his search for funds, George knocked on the doors of corporations, museums, and foundations. And he got used, if not immune, to a range of not-very-encouraging responses. There was the stuffy "That's not in our prescribed area of interest." There was the kiss-of-death "I'll get back to you." There was the pass-the-buck enthusiasm of "I've got a wonderful idea for you. Why don't you talk to the National Geographic Society?" Traveling to Washington, we *did* talk to the National Geographic Society. People there were intrigued by the project but ultimately decided it was "too dangerous." They had a point. If luck was with us, we would be the first archaeological team allowed into an area where the locals had for a number of years taken serious issue with the regime of Oman's Sultan Qaboos, expressing their discontent with Chinese-supplied machine guns and rocket launchers.

While in Washington, we rendezvoused with Barry Zorthian, a friend of George's who had once been a key CIA operative in Vietnam and now appeared to be working as a freelancer, though it was hard to tell for who or what (as it should be). Barry was genuinely fascinated by archaeology and thought he might be able to help us, for he had recently done some consulting for clients from the Sultanate of Oman.

Back in California, George's quest for funds produced Count Brando Crespi, scion of the French *Vogue* magazine empire. Meeting us for lunch at Baci, a suitably trendy restaurant, the Count wore a perfectly tailored off-white linen suit (but no socks) and spoke with an aristocratic lisp.

"I live in Milan, but keep pieds-à-terre in London and L.A.," he noted, then turned to the waiter to discuss the provenance of the porcini mushrooms cited on the menu.

"You hear that, Nick?" George whispered. "Pieds-à-terre here and there. Must have a pretty big foot."

He had long had an interest in archaeology, the count confided — *psychic* archaeology. He had, in fact, recently underwritten an extensive search for the tomb of Alexander the Great. Psychics in his employ had labored long and hard to visualize the final resting place of the Macedonian conqueror. They finally concluded that he lay buried beneath the Grand Mosque of Alexandria, an opinion shared by (and perhaps derived from) E. M. Forster's excellent 1947 *Guidebook to Alexandria.*

"But you just don't go into grand mosques and start ripping up the pavement," the Count lamented.

"No, I guess you don't," we sympathized.

The Count and his people had had no choice but to close down their Alexandria operation and go home. But perhaps his people could now help us locate Ubar. Yes? All they would need was a map and perhaps a few hints as to how they might focus their psychic powers.

"As an example, what kind of mosques did they have?" inquired the Count, mosques on his mind.

"They didn't have mosques, actually, back then," George gently explained. "Nick, maybe you could send Count Crespi a map?"

I did. Not just a single map of our search area, but a half-dozen randomly photocopied maps of different areas of Arabia, only one of which included our search area. And we heard nothing further from Count Crespi. At least he picked up the tab for lunch.

Over the course of the project, the Count wasn't the only one to volunteer psychic advice. One fellow placed Ubar on the banks of an antediluvian trans-Arabian canal. Another knew its whereabouts thanks to a direct line to Divine Revelation. And a "dowser into sacred geometry" offered us the services of "Jorsh," his spirit guide.

What more? Revelations from outer space? Actually, yes. Ron Blom called from the Jet Propulsion Lab. A magnetic tape had come in, containing the raw digital data of a Landsat 5 "Thematic Mapper" satellite pass over our search area. If we were lucky, a photographic image could be produced within a month.

"Any chance of a peek?" I wheedled. "Just a quick peek?"

"Let me see what I can do," Ron replied, ever good-natured.

The next afternoon, Ron cleared George Hedges and me through JPL security and led us to the Image Processing Lab. Its heart was a dark, hushed room in which not only the floor but the walls were thickly carpeted. At a dozen or so workstations, planetary scientists were caught in the glow of large color monitors. With a few taps on their keyboards, digital tapes rolled, were routed through a mainframe computer, and became beautifully rendered images of Earth and distant planets and moons.

Settling into our workstation, we were joined by data entry technician Jan Hayada and Bob Crippen, a colleague of Ron's who had a knack for matching wits with space images and coming up with unforeseen data. It was 3:15; we had until 4:00 to see what we could see.

Bob's fingers rippled across the keyboard. Nothing happened. "We're sharing a mainframe Data General," Ron noted. "Lot of people must be on it today. Slows it down." Then (slowly) a series of scans — cyan, magenta, yellow — swept down the screen and produced a high-resolution video image of our corner of far Arabia.

"This is a Landsat quarter-scene, sixty kilometers across. False color," Ron explained.

"Bands three, five, and nine," Bob added. "Original plus an eleven-by-eleven high-pass filter. Linear stretch. Standard stuff."

Ron nodded. As a result of "false color" imaging, the scene's gravel plains were rendered in unnatural blues and greens; dunes were painted rich shades of beige, ocher, and brown. We searched the scene: there wasn't a hint of anything man-made.

"Can we take a closer look? Here?" I wondered. Up and to the left was the outline of the ancient lakebed where Bertram Thomas had encountered "the road to Ubar."

"Sure," Bob replied, "It will take a few minutes, though." He centered a cross-hair cursor on the lakebed and typed in a sequence of commands.

We waited. Then, ever so slowly, the image rescanned. And we saw roads. Not one but several roads, coming from the east, crossing the lakebed, then branching off to both the west and the north.

"Fantastic!" George exclaimed.

"Not too shabby." Ron nodded and smiled. We were elated. But then it sank in that some, if not all, of these roads had to be modern, tracks laid down by oil prospectors, military patrols, and by free-wheeling bedouin of the age of the Toyota.

"Well, Bob . . ." Ron mused.

"Well, what?" Bob asked.

"Let's beat up on it," Ron said, a shade grimly.

"We could do an enhancement," Bob suggested, electronic gun-slinger eyes narrowing. "Divide one band by another, that sort of thing . . ."

A dense computer-tech conversation ensued, which I only vaguely followed. Ron theorized that ancient tracks — hammered by the feet of thousands of camels over hundreds of years — would be more compressed than those left by modern vehicles skittering over the terrain. Did this compression have a spectral signature?

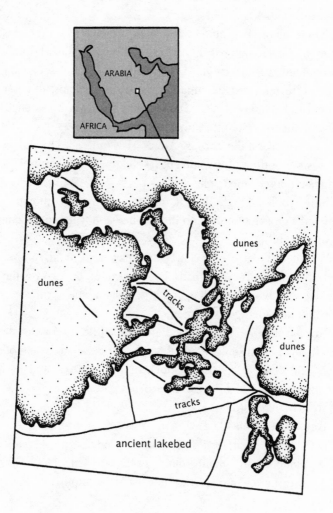

Detail of Landsat 5 image

"Hmm," Bob responded. He tentatively tapped his keyboard. "We could also hit up on the near-infrared, shift it to visible." Without waiting for a reply, Bob typed furiously. Then typed some more and rocked back in his chair.

"Hmm, we'll see."

The wall clock at the far end of the room read 3:52 P.M. The image rescanned, now producing waves of color that were surreal, psychedelic. And a strange and curious thing happened. The tracks crisscrossing the image faded away, disappeared. All but one.

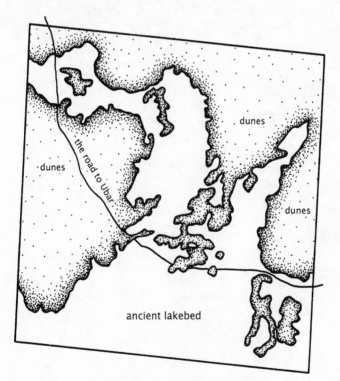

Landsat detail after image processing

"Well, look at that," said Bob. What remained on his monitor was a thin black line that arced up across the screen and led off into the dunes of the Rub' al-Khali.

We took turns, like little kids, following it with our fingers.

There it was, before our eyes and in far Arabia, the road to Ubar.

6

The Inscription of the Crows

IT DIDN'T TAKE LONG to come up with an expedition scenario. On the ground we would locate and follow the thin black line that angled across JPL's Landsat image. Somewhere on or near it would be Ubar — that is, if there ever was an Ubar.

In an enthusiastic letter, Father Jamme, the Jesuit epigrapher, gave us his blessing: "From JPL you have the revelation of an ancient road in the country of Ubar. Now you have a wonderful starting point which has to be discovered in place; it is an entirely new ball game. What a hope. Full speed ahead!"

Once JPL had made blow-ups of our satellite image, George Hedges framed one and sent it off to Ran Fiennes so that he could present it to Oman's Sultan Qaboos ibn Said, in the hope that he would eventually allow us to trek across his desert. Ran warned us, though, that we were not dealing with a land of quick responses.

Between us, Ron Blom and I spent hours scanning every millimeter of our full Landsat image and, tentatively, reconstructing the course of Oman's ancient Incense Road. It appeared that frankincense harvested in the mountains of Dhofar was first taken to two collection points, Hanun and Andhur, where there are known ruins. From Hanun and Andhur two separate routes went north across a gravel plain and converged at the well of Shisur.[1] From there a single route headed northwest and soon entered the dunes of the Rub' al-Khali. The telltale line on our Landsat image then shifted to the

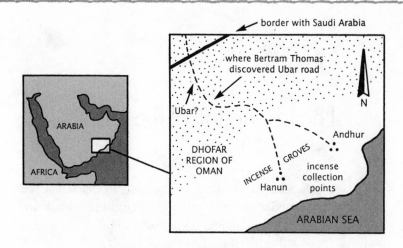

Incense caravan route through Oman

325-degree alignment reported by Bertram Thomas. As the dunes became ever more massive, the line became fainter and fainter. Just short of where Thomas thought Ubar was to be found, it all but vanished.

The road surely continued on, but we now could only approximate its route, for features generally have to be upward of thirty meters wide to show up on a Landsat image. For higher resolution, Ron suggested France's SPOT (Système Probatoire d'Observation de la Terre) satellite. Zeroing in on a smaller area and sacrificing color for black-and-white, it could image features less than ten meters across. Carefully, we plotted reference points for SPOT coverage so that as it arced over Arabia, the French satellite could align its lenses, prisms, and mirrors and further reveal the road to Ubar.

At about this time, we all agreed that we should recruit a trained, experienced archaeologist. It was one thing to look for Ubar, but if we found anything, any excavation would have to be directed by a professional. The obvious candidate for the job was Dr. Juris Zarins, who had recently completed a ten-year field survey of major archaeological sites in Saudi Arabia. Yet I hesitated to call him. His name conjured up the image of a long-bearded, stoop-shouldered academic, speaker of nine antique languages, who would say, dismissively, "Ubar! Ha! A fanciful myth, yes, but nothing more."

Zarins was currently teaching at Southwest Missouri State. When I finally phoned, I was surprised and delighted to find myself talking at great length with an immensely good-natured and enthusiastic individual. Within a week George Hedges and I were on a plane for Springfield, Missouri, to review the project.

It wasn't hard to break the ice with Juris Zarins — six foot four, mustache, but no long academic beard.

"Do you prefer to be called Juris or Juri?" George asked.

"Yes," he answered, deadpan, then broke into a broad smile and slapped us on our backs.

Juris/Juri had been born in Lithuania, raised in a German displaced persons camp, then had come to America with his family and settled in the Midwest. In high school he became a first-string basketball center, polished his English, and developed a lifelong fondness for terrible puns. And he took up the study of cuneiform.

"I guess the light struck early. I was interested in the origins of things, and even before college I got a summer job digging Lewis and Clark's fort. Guess what they gave me."

"The kitchen? Graveyard?" we ventured.

"The crapper. Personally excavated it."

With fellow farm kids, Juri served as a combat infantryman in Vietnam, then, on the GI Bill, earned a Ph.D. in archaeology from the University of Chicago. He was drawn to the Middle East and the evidence there of the origins of writing, cities, and civilization. Though he dearly loved his adopted Midwest, he thrived as well in the harsh, hostile deserts of the Mideast.

Within hours Juri was a member of our team. He thought Ubar could well be a real place. On the north side of the Rub' al-Khali, at the oasis of Jabrin, he had dug test pits that yielded bits of pottery from Mesopotamia. He knew that Mesopotamians had had trading colonies on the Arabian gulf coast, but what were they doing so far inland?

"It makes no sense," he mused, "unless, of course, there was an Ubar, and both sites functioned as stops for frankincense caravans." Heading north from Ubar, caravans could have followed a series of wells across the Rub' al-Khali to the oasis of Jabrin, then continued on their way to Mesopotamia. Zarins confirmed a larger vision of the Ubar road as a vital link in Arabia's incense trade (see map below).

By June of 1987 our now seven-member team had permission in the works, more space imagery on order, but still no money — so we had time to further research the myth of Ubar. Although I had more than

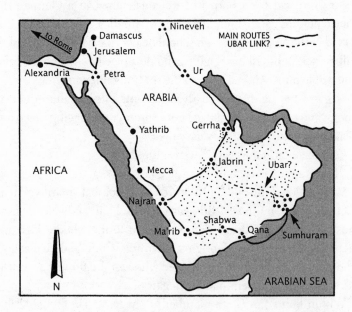

Arabia's incense roads

twenty loose-leaf binders of notes, I didn't know that there would be another twenty-seven to go. The more I read, the more there was to read in English and (with the help of translators) German, French, and Arabic — including modern Arabic, medieval Arabic, and Epigraphic South Arabic (ESA), a long unspoken language known only in inscriptions from the time of Ubar.

What archaeologists really treasure at their sites is anything in writing. Scrawled graffiti will do; graven inscriptions are better. The South Arabians had left behind an abundance of both, and yet, Father Jamme had assured me, none made mention of Ubar or its People of 'Ad. But now, in an obscure volume, I happened across the story of the *Palinuris*, a British survey ship that in 1824 charted the coastline of southern Arabia. At 54°30′E, 20°15′N, the *Palinuris* dropped anchor in a protected natural harbor. From its white sand beach rose a spectacular black volcanic hill, which the local tribesmen called Husn al-Ghurab, the Fortress of the Crows. Ascending to its summit, the ship's first mate, Lieutenant Wellstead, discovered an imposing, finely carved inscription. He diligently copied it down, and sent it on to the Reverend Charles Foster, an English cleric who pronounced it to be an 'Adite inscription of "awful antiquity." Foster's translation revealed it to be an account of the golden age and ultimate grief of the People of 'Ad! It even mentioned the prophet Hud, Ubar's doomsayer, by name. In part the long inscription read:

And we hunted the game, by land, with ropes and reeds;
And we drew forth the fishes from the depths of the sea.
Kings reigned over us, far removed from baseness,
And vehement against the people of perfidy and fraud.
They sanctioned for us, from the religion of Hud, right laws,
And we believed in miracles, the resurrection,
And the resurrection of the dead by the breath of God.

Then came years barren and burnt up:
When one evil year had passed away, there came another to
      succeed it.
And we became as though we had never seen a glimpse of
      good.

They died: and neither foot nor hoof remained.
Thus fares it with him who renders not thanks to God:
His footsteps fail not to be blotted out from his dwelling.[2]

What a wonderful poetic vision this was, except that it didn't quite make sense: early on in the inscription, its people "believed in the religion of Hud" and had kings "far removed from baseness," yet they later paid the price of "him who renders not thanks to God." And how could a people die off — "neither foot nor hoof" remaining — then pop back to elegize their own fate?

Ever hopeful, I sought Father Jamme's opinion, and by return mail received a freshly annotated translation. The inscription really did exist, he told me, but it had nothing to do with a lost golden age, the prophet Hud, or the promise of resurrection. The inscription was, in fact, the equivalent of the plaques that modern politicians delight in affixing to the walls of city halls and civic monuments. The inscription begins *not* with "And we hunted . . ." but with " S M Y F ' ' S W ' [a proper name] and his sons S R H B ' L Y K H L and M ' D R K B Y ' R [two more proper names] . . ." It then listed a city council of some forty additional names, all taking credit for renovating the fortifications at the Fortress of the Crows around 525 A.D.

Father Jamme advised that "C. Foster's contribution is best forgotten; its only value being that of a tiny historical dot." The good Jesuit signed off with a hearty "Cheerio!!!" So much for Lieutenant Wellstead's and the Reverend Mr. Foster's inscription. But give Foster credit, at least, for transforming a politician's plaque into a haunting myth. I imagined Charles Foster in a leaky-roofed Victorian parsonage, fitfully preparing a sermon for an indifferent congregation but secretly dreaming of an ancient people in a faraway sunny land, who once were favored by the grace of God.

What initially appealed to me about Foster's version of the inscription was that, had it been as he imagined, it would have bridged the divide between life today and life in the distant past. This, to me, was myth's great promise. Reading his wishful translation, you could fleetingly put yourself in the sandals of a people who "hunted the game, by land, with ropes and reeds" and yet endured "years barren and burned up" and suffered "as though we had never seen a glimpse of good." Though steeped in the fantastic, myth allows you to reach back to touch the lives of ancient people. To smell their spices, get the dust of their towns in your eyes. To dream, even, their dreams. Perhaps Foster felt the same way and was driven to invent what he could not find. His tale was cautionary.

Still, I couldn't help but feel that the Ubar myth offered a realistic promise. Already (I reassured myself) it had offered tangible clues as to the character and location of the site. And some cards remained to be turned over — in particular, the legend's grand flowering in Egypt and Persia in the 1100s. I looked forward to researching this period. There was the prospect that Ubar could convincingly become a real place . . . but it could also prove to be a city that never was, a concoction of medieval and ancient imagination.

The myth of Ubar, I was to find, had all the certainty of a desert mirage. It would draw you on, a vision of unexpected wonder rising from distant sands. Then suddenly, as you advanced just one step too far, it vanished. But then, as with a mirage, if you stepped back, the vision would return.

Were not mirages — despite their distortions and shimmering inversions — images of actual places, of palm trees and dwellings hidden beyond the curvature of the earth?

7

The Rawi's Tale

TO FURTHER EXPLORE the myth of Ubar, let us journey now
to Cairo in medieval times . . .

From across the desert, the traveler's view of the city's minarets and
domes is filtered by a silvery sepia haze rising from the kitchen fires of
palaces and hovels. Nearing Cairo and passing through the Bab
al-Futah, the Gate of Conquest, the traveler plunges into a dusky,
teeming labyrinth. Merchants crouched in tiny stalls cry out to a
stream of passersby to smell their spices and cinnamon, sample their
pomegranates and pistachio nuts. Down a crooked street, doctors
prescribe leeches for bile in the blood and horse oil for broken bones.
Caged birds twitter and shriek. A dark alley becomes so narrow that
two people can barely pass, then opens onto a square where ten
thousand souls praise Allah in the Great Mosque of ibn Tulun.

In the shade cast by the mosque's walls and pillars, *rawis*, itiner-
ant storytellers, practice their street theater. Their tales are by turns
bawdy, romantic, pious, and edged with suspense. The story of
Iram/Ubar is a favorite, oft-told and popular. Says one rawi, "Were I to
tell you of Iram's splendors and miraculous works, I fear you would
be calling me a liar, and by doing that you would be committing
a sin!"

By great good fortune, a sampling of rawis' tales of Iram/Ubar has
been preserved.[1] The most engaging is by a certain Muhammad ibn
Abdallah al-Kisai, of whom little is known. A guess would be that he

was a streetcorner rawi who at some point was adopted by the opulent court of the city's Abbasid pasha and thereby had the time and resources to commit his stories to writing.

In Cairo in 1180, we listen to his tale, which features the life and travails of Iram/Ubar's best-known figure and is therefore called the tale of . . .

The Prophet Hud

1   Know that in the beginning there were twelve male children of 'Ad son of Uz son of Aram son of Shem son of Noah, and God gave them power He has given to no one else.

I.   Hud and the Idolatrous People of 'Ad

Wahb ibn Munabbih [a prior chronicler] said: the greatest king of 'Ad was Khuljan; and he had three idols, Sada, Hird, and Haba, in the service of which he had placed one man for every day in the year. Among these, the noblest and best was Khulud. When this Khulud was asked why he had not married, since he 10   had reached the accustomed age, he replied, "Because in a dream I saw coming out of my loins a white chain, which had a light like the light of the sun. I heard a voice saying, 'Look well, Khulud, for when you see this chain come out of your loins again, marry the girl you will be commanded to marry.'" He was puzzled by this until one day he heard a voice say, "Khulud, marry the daughter of our uncle!" While he was asleep, suddenly the chain came forth from his loins.

When he awoke he went to his cousin, spoke for her, and was married to her. When he had lain with her, she conceived 20   Hud the prophet.

The ponds and rivers, the birds and beasts, wild and tame,

rejoiced at the conception of Hud. The trees of the tribe of 'Ad became green and brought forth fruit out of season by the blessing of Hud. And when his mother's days were accomplished, he was born on Friday.

One day, while he was at prayer, his mother saw him and asked, "My son, whom are you worshipping?"

"I am worshipping God, who created me and all creation," he answered.

30   "Do you not worship the idols?" asked his mother.

"Those idols bring neither harm nor profit," he said. "Neither do they see nor do they hear."

"My child," she said, "worship your God, for the day I conceived you I saw many strange things. When I was delivered of you in the valley, there were dry trees that became green and bore fruit. When I put you on a black rock, it became whiter than snow. Then I carried you home and saw a man whose head was in the sky and whose feet were in the vast expanses of the earth. He took you from me and raised you up to a people

40   in the sky whose faces were white. Then they returned you to me, and on your head were rays of light and on your arm was a green pearl. I heard one of them say, 'God has made you a prophet.' So act accordingly with what has appeared to you."

Kaab al-Ahbar [another chronicler] said: When Hud was four years old, God spoke to him, saying, "O Hud, I have selected you as a prophet and have made you a messenger of the tribe of 'Ad. Go therefore to them and fear them not. Call upon them to witness that there is no God but I alone, who have no partner, and that you are my servant and my messenger."

50   Hud went out to his people on the day of their great festival, held in the sandy regions called Ramal-alij [an old name for the

Rubʻ al-Khali]. Their king, Khuljan, was seated on a golden throne.

"*O my people,*" said Hud, "*worship God: ye have no other god than him*" (7.65).[2] So saying, he let out a great shout, and from afar the wild beasts and lions drew near and said, "We are at your service, O Hud. Inform us and have no fear."

But the hearts of the people were filled with fear; their faces turned pale, and they shuddered. They asked, "What are your
60 God's features like, his form, his length? Is he made of gold or silver?"

Hud described God's majesty. When he had finished his speech, the king said to him, "Do you think your Lord is more powerful than we are, considering the multitude of our numbers and the strength of our forces? Or do you not know that there are born to us every day and night one thousand two hundred male and female children?"

Hud replied: "*Did they not see that God, who had created them, was more mighty than they in strength?*" (41.15).
70 The first person to believe in Hud that day was one Junada, forty of whose cousins also believed. But the rest of the people rebuked and cursed Hud. He continued to indulge them, though, for a long time. Then God caused the women's wombs to become barren, and not a single woman among them bore a son or daughter. Hud never ceased warning them until he had been calling them to worship God for seventy years, but still they had no faith.

Kaab al-Ahbar said: Hud finally lifted his gaze to heaven and said, "O God, I ask thee to strike them down with famine and
80 drought. Perhaps then they will believe. If they do not, then I ask thee to destroy them through torment such as no one has been destroyed by before or will afterward."

So it was that God took away the rain and caused the earth to shrivel up, and no green thing grew in their fields and their beasts died; but they bore all this with patience for four years, until, despairing of themselves, they were about to believe. Thereupon King Khuljan told his subjects, "You must not enter into Hud's religion even if you be eating sand and drinking urine. If this suffering has afflicted us because of the multitude of our sins, why then have the wild beasts and animals of burden, which have no sin, been afflicted as much as we?"

Hud answered, calling out from a mountain-top, saying, "O children of 'Ad, if you have faith in your Lord, I will ask Him to send the heavens to you to pour down rain and to cause the earth to send forth her fruits."

II. *The Delegation to Mecca*

Ibn Abbas [yet another cited chronicler] said that in those days it was the custom, when a people was afflicted from heaven or from an enemy, to take offerings to the Sanctuary of the Ka'aba and to ask God for release from suffering. They would enter the Sanctuary mounted on she-camels adorned with diverse jewels.

In accordance with this custom, the 'Ad chose from among their nobles seventy men. Seven were chosen as leaders, and their names were Qayl, Luqman, Jahlama, Ubayl, Marthid (who believed in Hud), Amr, and Luqaym. While they were departing from their land, they heard a voice saying, "Despair and misery for you, O House of 'Ad! You shall perish, and a destructive, shifting, icy gale, turbulent with dust, will descend upon you." They paid no attention to the voice, however, and went along their way.

When the delegation arrived seeking entrance into the Sanctuary, they heard a voice saying,

"May God vanquish the delegation of 'Ad:
They have traveled to pray for rain;
May they quench their thirst with hot water!"

The king of Mecca at that time was called Muawiya ibn Bakr.
The delegation descended on his house and remained there for
one month, eating and drinking, and forgot what they had
come for. But Muawiya was loathe to ask them to leave his
120 house, although it was said that all this hospitality had grown
burdensome for him. Therefore, he sent them two slave-girls,
called the Two Locusts, who were singers in his service. He said
to them, "While they are eating and drinking, sing to them and
make them desirous of praying for rain." The two girls sang,

"Woe unto you! Woe unto 'Ad!
Because of great thirst neither grand lord nor slave
has hope.
O delegation of drunks, remember your tribe,
parched with thirst."

130 When they heard what the slave-girls said, they bathed
themselves, put on clothes not soiled by wine, approached the
Sanctuary, and draped it with their robes; but the Sanctuary
would not accept them.

One of the men said, "Shall we abandon the religion of our
noble and meritorious fathers, and follow the religion of Hud?"

"O God," said Marthid (who believed in Hud), "you are
right to send torment to those who believe not!"

III. *God's Vengeance*

God commanded the angel of the clouds to spread over them
140 three clouds, one white, one red, and one black. When the del-

egation returning from Mecca saw these clouds, they rejoiced. But one of them was ordered, "O Qayl, choose for your people one of these three clouds!" He chose the black one and was told, "O Qayl, you have chosen the black cloud, in which are ashes and lead. 'Ad shall perish to the last from the heat!"

The cloud moved until it had emerged from Wadi al-Mughith. When the people of 'Ad saw it, they said, "This cloud has come to give us rain!"

God's angel Gabriel said, "O cloud of the Barren Wind, be a
150 torment to the people of 'Ad and a mercy to others!"

On the first day the wind came so cold and gray that it left nothing on the face of the earth unshattered. On the second day there was a yellow wind that touched nothing it did not tear up and throw into the air. On the third day a red wind left nothing undestroyed. And the wind kept on blowing over them for eight unhappy days and seven hapless nights. On the eighth day the 'Adites lined up and began to shoot arrows at the wind, saying, "We are mightier than you, Lord of Hud!"

Thereupon the wind ripped them apart and went into their
160 clothing, raised them into the air and cast them down on their heads, dead. The wind snatched their arrows and drove them into their throats. Thus it continued until there was left of them only their king, who remained to be shown what had become of his people. He fended the wind with his chest and said, "Woe on this terrible day! Sons and thrones are destroyed!"

Then the wind entered his mouth and came out his posterior, and he fell down dead. The wind hurled the palaces together and killed all the women and children that were in them. It passed on to the sanctuary and raised them into the air
170 and cast them down on their heads, dead. As God hath said: *And when our sentence came to be put in execution, we delivered Hud, and those who had believed in him, through our mercy* (11.58).

Hud and those believers who were with him traveled to the Yemen, where they camped. They remained there for two full years, then death took him and he was buried in the Hadramaut.

Kaab al-Ahbar said: One day I was in the Prophet's Mosque during the caliphate of Othman. A man entered the mosque, 180 and everybody stared at him because of his height.

"I am from the Hadramaut," he said, and he spoke of Hud's grave.

"In my youth I went with a group of lads of my own people, and we traveled through the land of the sandy desert until we reached a high mountain, where in a cave we found a huge rock stacked on top of another rock, and between the two was an opening through which only a thin man could pass. As I was the tiniest of the group, I entered and found a throne of red gold on which sat a dead man. I touched his body; he was Hud. 190 I looked at him and saw that his eyes were large and his eyebrows met. He had a wide forehead, an oval face, fine feet, and a long beard. Over his head was a rock shaped like a board, on which were written three lines in Indian letters. The first of these said, 'There is no god but God; Muhammad is God's messenger.' On the second was written, 'I am Hud ibn Khulud ibn Saad ibn 'Ad, God's apostle to the tribe of 'Ad. I came to them with the message, and they denied me. God took them with the Barren Wind.'"

This tale told in dusky medieval Cairo illustrates why the Ubar myth has survived for many a century. It is a good yarn, here related by a skilled and stirring storyteller.

On the lookout for Ubar clues, I was first intrigued by the choice of three clouds offered to the tale's rapidly sobering delegation of drunks (lines 140–145). I was aware that a three-way choice was a venerable Semitic theme; similar choices are described in the Bible and in accounts of Arabian soothsaying. But why are the three clouds white, then red, then black? The answer came in a flash of perception from JPL's Ron Blom.

"Tell you what it sounds like to me," he remarked over lunch at the lab's cafeteria, "sounds like a report of a volcanic eruption. First there's a cloud of white smoke, then comes a rain of red magma, and finally black ash falls. Like the old story says, 'ashes and lead.' But I don't recall any volcanoes where we're looking for Ubar."

There weren't any. Nor were three clouds mentioned in the Koran, the earliest coherent record of the Iram/Ubar story. What likely happened is that a report of an explosive volcanic eruption — possibly Vesuvius in 79 or 512 A.D. — was rung in for its dramatic value, as an effective way to set up the city's destruction.

As I further studied the tale, I found that the three clouds weren't the only elements slipped in after the fact. The running description of the travails of the prophet Hud, for instance, turned out to be based on the considerably later and unconnected experiences of the prophet Muhammad as he preached a new religion in Mecca and was spurned by his tribe.

It became evident that this tale of al-Kisai, on the surface relatively straightforward, had a complex subtext, highly symbolic and replete with allusions-within-allusions. (For a look at this subtext, see the Appendix, page 280.) Essentially, much of the tale is immensely intriguing but has little or nothing to do with Ubar as a real place. On the other hand, "The Prophet Hud" incorporates lore that may indeed bear on an actual city's rise and fall. The tale offers potential insights into Ubar's antiquity, its people, its destruction, and its location.

ANTIQUITY   Al-Kisai's tale is heralded by a genealogy (lines 1–3). Arab storytellers loved genealogies, for they imparted a ring of authenticity. They also harked back to a time before writing, when an individual and his tribe were defined by their place in a long and worthy procession of remembered ancestors. What is unusual is that by Arab standards, this genealogy's line of descent is remarkably short, with only a half-dozen generations between Noah and the glory days of Iram/Ubar. This would make the People of 'Ad an ancient tribe, perhaps the oldest in all Arabia.

PEOPLE   The People of 'Ad appear to have had a lively appreciation of sin, though the nature and extent of their sinfulness is unclear. They were certainly materialistic, and they worshipped at least three gods. Confronting them, the prophet Hud made quite an impact preaching the worship of but a single God. Who, then, was Hud?

In Semitic lore (shared by both Jews and Arabs), names frequently have an elemental allegorical meaning: for Daoud, or David, it is "beloved"; for Suleiman, or Solomon, it is "man of peace." And "Hud" comes from the root HWD: "to be Jewish." This linkage is clearly reflected in the Arabic of the Koran, where "Hud" is not just the name of a prophet but a collective noun denoting the Jews.

Was Hud Jewish?

He could well have been. Much has been written of the Jews of Arabia, much of it chronicling the period 300–525. In that era Jewish courtiers and even a Jewish king ruled the kingdom of Himyar, which rose and fell in what today is Yemen. It is no stretch of the imagination to believe that a Jewish trader — or even a rabbi — could have made his way to Ubar and preached the religion of a single God.

Was Hud a historical figure? It is impossible to know. But he certainly was a compelling *allegorical* figure, a symbol of early monotheism. The monotheism of Arabia and Islam, the prophet Muhammad himself declared, was heir to revelations first made to the Jews.

DESTRUCTION   Iram/Ubar was suddenly destroyed by a great cataclysm. But what actually happened? The storyteller appears unsure of his apocalyptic imagery. Early on, a mysterious voice promises destruction by an "icy gale, turbulent with dust" (line 108). A few sentences later another voice exclaims, "May they [the 'Adites] quench their thirst with hot water" (line 115). Next we are given the imagery of the three volcanic clouds. The Barren Wind that is ultimately unleashed is supremely violent but not terribly convincing. It has the characteristics of a tornado, an all-but-unknown phenomenon in southern Arabia.

Other versions of the tale downplay the wind, having Iram/Ubar destroyed by a mighty but still baffling "Divine Shout." In one account the end comes as "suddenly the earth opened around it and Iram, bathed in a strange twilight, began to sink slowly down until the whole city was completely swallowed up. All that remained was an endless wilderness of empty, shifting sands across which the winds moaned and howled."[3]

If we ever did find Iram/Ubar, it certainly would be interesting to investigate how (or even if) the settlement was destroyed.

LOCATION The casual mention of the "Wadi al-Mughith" (line 147) was to provide a valuable clue to Ubar's location. At first the name didn't make any sense; it's not to be found on any map, old or new. But then, by chance, I found it cited in an account by the 800s A.D. geographer Ibn Sa'd, except that he expands the name to the "Wadi al-Mughith *in Sihr.*" This rang a bell from my earlier Ubar research. The word "Sihr" appears on something called "The Gardens of Humanity and the Amusement of the Soul," a map compiled by Muhammad al-Idrisi, an Arab cartographer living in Sicily in the 1100s A.D. Though fanciful in title, the map is an outstanding example of the Arab scholarship that through the Dark Ages kept a flickering flame of learning alive. Based on earlier records and the accounts of mariners and merchants, it depicts Arabia in detail. The Ubar region bears the legend "the region of 'Ad and the place of *Sihr.*"

Here, then, was an intriguing two-step connection: our rawi's tale placed Ubar near the Wadi al-Mughith, which another writer called the Wadi al-Mughith in *Sihr.* His addition of the word "Sihr" leads us

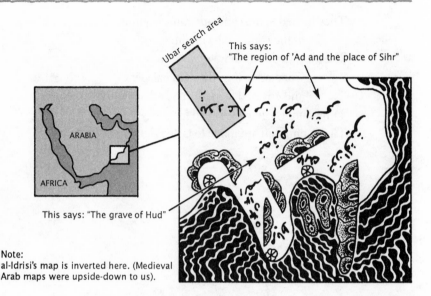

Detail of al-Idrisi's map of Arabia

to a locale on a respected early map of Arabia, a detail of which is shown here. The shaded box approximates what we had tentatively settled on as our Ubar search area. We were, it appeared, in the right neighborhood!

Several worthies of storyteller al-Kisai's time locate Iram/Ubar and the People of 'Ad in the same patch of Arabia. The historian Nashwan ibn Sa'id al-Himyari (died 1117) sums up: "Ubar is . . . the name of the land which belonged to 'Ad in the eastern part of Yemen; today it is an untrodden desert owing to the drying up of its water. There are to be found in it great buildings which the wind has smothered in sand." (At the time, eastern Yemen included part of what today is Oman.)[4]

This Nashwan al-Himyari was also a poet, who dwelt on the theme of vainglory come to grief. His verses are an epitaph for Iram/Ubar and the People of 'Ad:

They turned to dust and are trampled under foot,
      as they once did with others. . .
They rest in the earth now,
      when once they dwelt in palaces
      and enjoyed food, drink, and beautiful women.
Time mingles good with bad fortune,
Time's children are made to taste grief amidst joy.[5]

8

Should You Eat Something That Talks to You?

THE APPROACH I HAD TAKEN with al-Kisai's tale "The Prophet Hud" was to look closely at the story, identify the elements added on over the years, and throw them out, hoping that maybe — always maybe — something would be left of a real time and place. This approach also worked well with the story of the geographer Yaqut ibn 'Abdallah (died 1229). "Yaqut" was a slave's name meaning "Ruby." In the medieval Arab world, it was common to give slaves names of gems and flowers and virtues, allowing their master to say, "Bring me dates, Pleasure. Tell me a story, Ruby."

As a trusted slave, Yaqut traveled the Persian Gulf on behalf of his master, a Baghdad merchant. He increased his master's prosperity, traveled farther, and eventually was given leave to compile his *Mujam al-Buldam*, or "Dictionary of Lands." It includes a substantial section on the "City of Wabar," or Ubar. Featuring lore Yaqut gathered while trading in Oman, most of his description is fanciful. Yaqut was particularly taken by the notion of nisnases. He wrote, "The Lord destroyed everything there [in Ubar] and converted the human beings to nisnas — a monkey-like creature with a human-being look. The men and women appeared ridiculous, each had half a head and half a face with only one eye and one arm and one leg . . . They used to jump high and rapidly on that one leg. God made them chew up the grass like cows and buffalos."[1]

Yaqut gives us anecdotal accounts of bedouin run-ins with "devil man" nisnases — and he is disquieted by the fact that nisnases speak Arabic. This, he reports, troubled the bedouin as well, who felt it was okay to hunt down and spear a nisnas (just as they would slay any other creature), but, they asked, should you eat something that talked to you?

Because of his preoccupation with nisnases, both Arabic and Western scholars have ridiculed and discounted Yaqut's account of Ubar. But subtract all the nisnas lore, and what's left is an intriguing geographical description: "Wabar is a vast piece of land, about 300 fersakh [37½ miles] wide. . . . The land of Wabar was very much fertile and very rich with water. It was full of trees and fruits. The very fast growing population there could multiply their wealth and could live in excessive luxury. . . . [There is] a big well called the Well of Wabar."[2]

Here we have a thumbnail sketch of Ubar as it really might have been: a sizable oasis watered by a great well, and the home of a rapidly expanding, prosperous population that became a little too full of themselves.

9

The City of Brass

As the tale of Ubar was told and retold throughout medieval times, any clues to its underlying reality became more and more submerged in fantasy. This is evident in the variations of the legend that appear in *Alf Laylah wah Layla*, the *Arabian Nights*.

The *Arabian Nights* tales have long been considered the product of overheated imagination. When they were popularized in English at the turn of our century, essayist Thomas Carlyle considered them "unwholesome literature" and forbade their presence in his house. "Downright lies," he sniffed. "No sober stone is permitted to kill even the wildest fantasies." Even back in the 900s, historian Abu al-Hasan Ali al-Mas'udi termed the tales "vulgar, insipid"; nevertheless, he wrote of their origin: "The first who composed tales and made books of them were the Persians. The Arabs translated them and the learned took them and embellished them and composed others like them."[1]

Still relatively unexplored is a possible *pre*-Persian origin of the tales. Ethnologist Leo Frobenius and mythologist Joseph Campbell both suspected that a number of the tales were in their earliest forms genuinely Arabian — and had their genesis in the valley of the Hadramaut, not far west of our search area.[2] (It was in this valley that Ubar's Prophet Hud was traditionally said to lie buried.) Given this proximity — and given that Iram/Ubar's People of 'Ad wander in and out of many of the tales in the *Arabian Nights* — I thought they might conceal a wealth of useful information.

A succinct, relatively restrained tale is "The Story of Many Columned Iram and Abdallah Son of Abi Kilabah." As for clues to Ubar, there is an interesting description of the city's locale as "an uninhabited spot, a vast and fair open plain clear of sand-hills and mountains, with founts flushing . . ." And, following Ubar's divine demolition, there is the line "Moreover, Allah blotted out the road which led to the city" — a reference, it would appear, to the very road we were seeking.[3]

In other *Arabian Nights* tales, the myth of Ubar gathers momentum — and fancy — as it careens from story to story. In "The Eldest Lady's Tale" a medieval traveler enters a version of Ubar and finds it quite intact. It has not been destroyed; rather, its inhabitants "had been translated by the anger of Allah and had become stones . . . all were into black stones enstoned: not an inhabited house appeared to the espier, nor a blower of fire. We were awe struck at the sight . . . and said 'Doubtless there is some mystery in all this.'"[4]

In this version the Iram/Ubar myth has been populated by *maskoot*, an Arabic term for human beings petrified by the wrath of God. It has been suggested that the idea of maskoot came from the discovery by superstitious bedouin of shattered statuary in the deserts of Upper Egypt.

More maskoot are found in the tale "The City of Brass," by all odds the most bizarre rendition of the Ubar myth. It is phantasmagoric, funereal. It recounts the adventures of Emir Musa and his companions as they set out from the sea in search of a mysterious lost city. They are directed across a great desert by a brass horseman, then by a djinn (or genie) mired up to his armpits in a furnace. The way "is full of dozens of frights, full of wonders and strange things." Emir Musa's little group finally climbs a hill, and . . .

When they reached the top, they beheld beneath them a city, never saw eyes a greater or goodlier, with dwelling-places and mansions of towering height, and palaces and pavilions and domes gleaming gloriously bright . . . and its streams were a-flowing and flowers a-blowing and fruits a-glowing. It was a city with gates impregnable; but void and still, without a voice or a cheering inhabitant. The owl hooted in its quarters; the bird skimmed circling over its squares and the raven croaked in its great thoroughfares weeping and be-wailing the dwellers who erst made it their dwelling.[5]

Within its walls, the city is a showcase of death, at once splendid and the stuff of nightmares. Its streets and palaces are filled with ghastly long-dead maskoot. The queen of Sheba even makes a cameo appearance. Reclining on a bejeweled couch, she appears as "the luce-dent sun, eyes never saw a fairer." But . . . "she is a corpse embalmed with exceeding art; her eyes were taken out after her death and quicksilver set under them, after which they were restored to their sockets. Wherefore they glisten and when the air moveth the lashes, she seemed to wink and it appeareth to the beholder as though she looked at him."

Everywhere in this city of the petrified are graven inscriptions that drive home the moral message: *the pleasures of immense riches — here everywhere in evidence — are for naught, for life is brief and death almighty.* This powerful idea is central to medieval Islam. As a legend over the tomb of the son of King Shaddad ibn 'Ad tells us: "Be warned by my example. I amassed treasures beyond the competence of all the kings of the earth, deeming that delight would still endure to me. But there fell on me unawares the Destroyer of delights and the Sunderer of societies, the Desolator of domiciles and the Spoiler of inhabited spots." The angel of Death.

Though its message and central story line are relatively simple, "The City of Brass" in its entirety is a dense, complex tale, a concate-nation of imagery and characters from diverse times and places. Emir Musa appears to be an incarnation of the Old Testament's Elijah.

And what are not only the queen of Sheba but Solomon and Alexander the Great doing here? The tale, packed with hundreds of allusions, is based on at least eight major sources. Indeed, the brass city itself appears to be inspired not only by Ubar but by rumors of a town of copper and brass located either in North Africa or in Spain. To accommodate these rumors, whoever wrote the tale whisked Ubar and its People of 'Ad from Arabia and plunked them down in Andalusia!

"The City of Brass" is ultimately surreal, beckoning us ever onward into a sun-drenched yet sinister landscape that "is like unto the dreams of the dreamer and the sleep-visions of the sleeper or as the mirage of the desert, which the thirsty take for water; and Satan maketh it fair for men even unto death."

The tale draws to a close with hardly a glimmer of Ubar as a real place. If mythmaking can be likened to a mirage — hiding yet reflecting a distant reality — that mirage has finally become all but impenetrable. Emir Musa and his companions take leave of the sepulchral brass city they have discovered and retrace their steps across the desert to the sea. But then, in the span of a single sentence, the mirage fleetingly dissolves. We are told that the emir's party "came in sight of a high mountain overlooking the sea and full of caves, wherein dwelt a tribe of blacks, clad in hides, with burnooses also of hide and speaking an unknown tongue."

Here, quite unexpectedly, is a sudden cluster of clues.

1. "a high mountain overlooking the sea and full of caves . . ." In all of Arabia, the *only* seaside peaks known for their caves are the Dhofar Mountains in southern Oman.
2. "caves, wherein dwelt a tribe . . ." The peninsula's only cave dwellers — past or present — are tribes of the Dhofar Mountains.
3. "a tribe of blacks . . ." The people of the Dhofar Mountains are distinctly dark-skinned.

4. "clad in hides, with burnooses also of hide . . ." Hides come from cattle, and there is only one area of Arabia where within the past 4,000 years cattle have been a mainstay of a tribe's livelihood: the Dhofar Mountains.
5. "and speaking an unknown tongue." The Dhofar Mountains are the *only* area where a language other than Arabic is spoken — an ancient language, only recently studied.[6]

Were these legitimate clues? Or was this a cluster of incredible coincidences? In any case, if we ever found the wherewithal to search for Ubar, we would begin our journey *at the foot of the Dhofar Mountains.* We would cross these mountains, where to this day dark-skinned people raise cattle and speak a strange language. Until recently, they dwelt in caves.

This was the logical route to our search area in the desert beyond. In following it, we would be setting out in the footsteps of Bertram Thomas, Wendell Phillips — and now, apparently, Emir Musa of the *Arabian Nights.*

10

The Singing Sands

AT THE TURN OF 1990 I was at work, alone, sorting out paperwork for a documentary film for Occidental Petroleum. I gazed out the window of my office at the company's Los Angeles headquarters, and instead of steel and glass towers saw the sands of Arabia. Was Ubar really out there? Or was it only a city of the imagination — as real as brass horsemen, djinns in furnaces, and queens with quicksilver eyes.

The phone rang. It was George Hedges. "We got a letter from Yahya," he said. His voice was curiously flat, as if he was feigning nonchalance.

Yahya. Another mystic? Like the Count, like Jorsh?

"Yahya . . ." George repeated. "According to his letterhead, he's with the Oman International Bank. They want to sponsor us."

George explained that Barry Zorthian, the ex-CIA operative we had met in Washington, had brought our project to the attention of Dr. Omar Zawawi, chairman of the Oman International Bank. A physician and philanthropist as well as entrepreneur, Dr. Zawawi liked the idea of looking for Ubar and asked his associate Yahya Abdullah to contact us. If we could make a brief preliminary trip to Oman, Dr. Zawawi's bank would not only pay our way but help us enlist additional sponsors, who would either help underwrite the expedition or donate services and equipment. At the same time we could get a feel for the problems we would face and perhaps even fit in a quick reconnaissance of our search area.

"How about that?" said George with a deep and heartfelt sigh. "At last!"

We called Juri Zarins. He was delighted with the news. At the time he was researching the demand for incense in the ancient kingdoms of Mesopotamia, a need that may have been met by shipments from Ubar.

We called Ran Fiennes. With the influential Dr. Zawawi behind us, Ran felt that permission would be granted for at least an Ubar reconnaissance. There was a hitch, though. Ran was about to set out on another Arctic adventure — a walk, unassisted by machine or dog, to the North Pole. The earliest he could fit in the Ubar reconnaissance was the following summer. We settled on the last two weeks in July. Considering that nearly ten years had gone by, what was another few months? The delay, in fact, would give us time to analyze some space imaging that was due in, in fact overdue, from the French SPOT satellite.

At the Jet Propulsion Lab, Ron Blom and Charles Elachi were delighted that an expedition was now possible. "But," Ron grumbled, "I don't get it."

"Get what?" I asked.

"What's wrong with the French?" What Ron had in mind was their inexplicable delay in forwarding computer tapes of the SPOT satellite's pass over our search area. It was not as if JPL/NASA had written a bad check.

Ron called the SPOT people and got the verbal equivalent of an exasperated shrug. The reason for the delay, they explained, was that the satellite had twice overflown our search area, and twice sent back unusable images. Too many clouds.

Clouds in the Rub' al-Khali? Unlikely, Ron thought. And then he recalled that the French routinely spot-checked incoming images with low-resolution, low-quality scans. "Perhaps you're not seeing clouds at all," he suggested. "Perhaps those clouds are dunes."

They were. And a month later we had images that were worth the

wait. In black-and-white rather than color, they had triple the resolution of our previous Landsat 5 shots. The road to Ubar was razor
sharp and clearly visible as it ran far out into the dunes of the Rub'
al-Khali.

As I studied the image, absorbed in what might lie along the Ubar
road, Ron and Bob Crippin whispered back and forth; I caught
phrases like "computer waypoints" and "pixel registration."

"We can do better," they announced.

What they had in mind was a computer-generated merge of data
from our Landsat 5 and SPOT images. The result would be a single
image that had the sharpness of the black-and-white SPOT data *and*
the rich color of the Landsat 5 imagery. It was a technologically
daunting idea. Two disparate pictures, taken years apart from different altitudes and angles, with different lenses, would have to be
precisely overlaid: 36,000,000 pixels of SPOT information superimposed on 16,262,000 pixels of Landsat 5 information.

It worked, and the merged image was detailed and dazzling. If the
expedition did go forward, we had plenty of candidates for Ubar to
check out. The most promising was one we named the "L" site. Our
"road to Ubar" led to a sharply defined L (400 by 400 meters) that
appeared to be man-made. It could be an agricultural area — or even
a walled, ruined city. There was nothing remotely like it anywhere
else on our images.

In early May I compiled a list of coordinates of points of interest
along our road and faxed them to Ran, who in turn forwarded them
to the Omani military authorities. There was a chance that we could
have the use of a military aircraft for our reconnaissance. If so, we had
a flight plan.

At about this time a very curious thing happened. Finding myself
with an unexpected few days off, I decided to take a break not only
from filmmaking but from the Ubar project. I was getting a little
obsessive about it, to say the least. In our garage I dusted off my

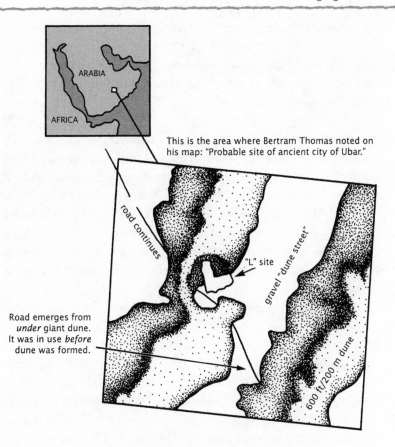

This is the area where Bertram Thomas noted on his map: "Probable site of ancient city of Ubar."

ARABIA

AFRICA

road continues

"L" site

gravel "dune street"

Road emerges from *under* giant dune. It was in use *before* dune was formed.

600 ft/200 m dune

Landsat 5 / SPOT composite image of the Ubar road

thirty-year-old Raleigh bicycle. Though a clunker by current standards, it had in recent years taken me on longer and longer solos out across the deserts of the Southwest. What could be better than a swing out across the Mojave, then through Joshua Tree National Monument, and on into my favorite desert, the Anza-Borrego? It would be good exercise. I'd enjoy clear air, sweeping scenery, and, for company, a couple of paperback mysteries. My daughter Jennifer recommended I take something by the English writer Josephine Tey. I picked *The Singing Sands*, which had a rod, reel, and a trout on the cover; it appeared to be a tale of fishing and felony in damp, dull-skied Scotland.

And it was . . . until my Raleigh and I stopped for Gatorade and pretzels (lunch when I'm left to my own devices) in the shade of a sandstone outcropping. Inspector Hugh Grant, Tey's Scotland Yard detective, has been puzzling over a murder on the London-Aberdeen sleeper. The victim, Grant learns, had been a pilot for Orient Commercial Airlines, an outfit that ran freight to southern Arabia. Grant conjectures that somewhere in Arabia the pilot may have been driven off course by a windstorm — and from the air discovered something incredibly rare and strange, something that led to his death.

I cycled on through the heat of the afternoon and thought about *The Singing Sands*. What was Arabia doing in this story? What was the something the pilot saw, the something worth killing for?

I stopped again for Gatorade and a few more pages. In Scotland, Inspector Grant drops by a local library to read up on Arabia — and happens on a description of a place called Wabar: "Wabar, it seemed, was the Atlantis of Arabia. The fabled city of Ad ibn Kin'ad. Sometime in the time between legend and history it had been destroyed by fire for its sins. . . . And now Wabar, the fabled city, was a cluster of ruins guarded by the shifting sands, by cliffs of stone that forever changed place and form; and inhabited by a monkey race and by evil jinns."[1]

Ubar! I sank down by the edge of the road and read on. A character based on Harry St. John Philby or Wilfred Thesiger, it is hard to tell which, is drawn into the plot. I suspected that one of the two men had rubbed Josephine Tey very much the wrong way, for the character is querulous, effete, a creature of "pathological vanity." Had Harry or

Wilfred snubbed Josephine? And now, with this mystery, was she having her revenge?

As the sun fell lower in the sky, I wondered if perhaps Ubar had already been discovered and *The Singing Sands* was a fictionalization of what had happened! Although the Philby-Thesiger character doesn't find Ubar, someone else does. Sitting down to his coffee and scones, Inspector Grant opens the morning edition of the London *Clarion* and is startled by the headline "SHANGRI-LA REALLY EXISTS. SENSATIONAL DISCOVERY. HISTORIC FIND IN ARABIA."

He turns to the *Morning News*, which confirms "ASTOUNDING NEWS FROM ARABIA."

It was dark now, and chilly where I had stopped. I was twenty miles from anywhere. But I read on by flashlight, grimly determined to see how *The Singing Sands* came out. As the moon rose and shone upon the desert, I was relieved to find that Inspector Grant cracks the case, and the Philby-Thesiger character gets his comeuppance. And it became clear that Ubar was still out there, waiting to be found. The city's discovery in the book happened only in the writer's imagination. For a long afternoon and a good part of the evening, though, she had me fooled.

I pedaled on. It was cold, cold enough, fortunately, for the rattlesnakes to stay in their burrows rather than stretch out on the blacktop for warmth. I thought over *The Singing Sands*. It was clever, well researched, and encouraging. Midway through the story Inspector Grant observes, "None of the writers [that he had consulted] attempted to belittle or discount the legend.... The story was universal in Arabia and constant in its form, and sentimentalist and scientist alike believed that it had its basis in fact . . . but the sands and the jinns and the mirages had guarded it well."

Ahead now were the lights of the little town of Borrego Springs where Kay and our daughters were to join me.

On the weekend, we would hike the desert washes and afterward share some Cerveza Pacificas with the park rangers who patrol the Anza-Borrego region. One, naturalist Mark Jorgensen, had spent some time in Arabia. He told us, "When you get out there, be sure to drink water, water, water. More than you would in our desert. Don't ration yourself. The human body is designed to operate at temperatures of up to 130 degrees, provided it has enough water."

It would prove to be good advice.

II

Expedition

11

Reconnaissance

IN THE SULTANATE OF OMAN on an August morning in 1990, the overnight Gulf Air flight from London rolled to a stop. Aboard was our team: Kay and I, George Hedges, Ran Fiennes, Ron Blom, and Juri Zarins. As the plane's door swung open, our impression of Muscat was, quite literally, a blur. Our eyeglasses were instantaneously fogged by the 100-percent humidity and the 120-degree heat, in the shade.

When we could see again, out the window of an air-conditioned van, it was clear that in the ten years since Kay and I had been here, Muscat had boomed. Everywhere we saw new buildings, lush landscaping, and extraordinary municipal monuments. To our right, a heroic hand burst from the shrubbery and thrust an even more heroic sword a good fifty feet into the air. To the left, a herd of oversize fiberglass oryxes placidly grazed. Down the road, a giant incense burner smoked by day and blazed with lasers by night.

This splendor, we learned, was at the behest of H.M. Sultan Qaboos ibn Said, a not only benevolent but imaginative absolute ruler. He was proud of his country's heritage. Indeed, he had recently issued edicts that Omanis should wear only traditional dress and that the colors of buildings should conform with a personally selected (and quite pleasing) palette of traditional hues. But H.M., as everyone knew him, was no isolationist; he admired the heritage of the West as well as the East. He loved Bach, had a palace organist (and organ), and had decreed that before noon only classical music should be played on the radio.

Oil exploration had favored the sultan and the Omanis, and overall the proceeds appeared to have been wisely spent on first-class roads, schools, public health (you could drink the water anywhere), and hospitals. The country appeared serene, though there were hints that its current prosperity had suspended but not dissolved tribal rivalries and that old religious enmities — some having to do with Islam and the West — were not far from the surface. Early on, a deputy minister did, quite diplomatically, question whether it was proper for us, as westerners, to seek Ubar.

"There might be some people — not me, of course — who might have an objection."

"To?"

"Looking for a city that is in the Koran."

"I see . . . But these people, the ones at Ubar. Doesn't the Koran say they were bad people? Like, wicked?"

"True . . ."

"So Ubar couldn't be a holy city. In fact, it would have been anything but a holy city. It would be . . ."

"Sin city!" the deputy volunteered. "Quite. Well, there you are."

We were mutually relieved, though I sensed the deputy was not altogether convinced by this line of reasoning. Always, our team decided, we should be as circumspect as possible when it came to religious matters. As a result, more than once we would find ourselves faux-piously sipping orange juice as Omani hosts knocked down gin and tonics.

Most of that week was taken up by a round of meetings in Muscat orchestrated by Malik al-Hinai, of late an officer of the sultan's Palace Guard and now with the Oman International Bank, our initial sponsor. We paid calls on ministries. We made appeals to potential sponsors. Silver-tongued and charming, Sir Ranulph was a master at this. He initially estimated that a proper expedition would cost $35,000, then arbitrarily upped the figure to $78,000. When no one seemed to blink, he further raised the ante. "What's he saying? Where did he get that?" George Hedges whispered when, in the midst of a presentation, Ran offhandedly remarked that $180,000 should see us through.

Though we eventually did raise a modest amount of cash, our wherewithal to look for Ubar proved to be in-kind donations by companies from a variety of countries. Gulf Air offered to fly us between England and Oman. While in Muscat we would be put up at the al-Bustan Palace Hotel. Out in the desert, "the Official Vehicle of the Ubar Expedition" would be the Land Rover Discovery. To keep in touch we would use French Racal radios, and we would log our finds on IBM computers. And we most gratefully accepted Scotland's Rowntree-Mackintosh as the expedition's exclusive chocolatier. Their Kit Kat bars became an expedition staple. At one point I had occasion to request a favor from a desert imam. Ran translated the religious leader's reply: "He says just give him Kit Kats, and it's anything we want."

By the end of the week, all was well in Muscat and we were winging south in a single-engine World War II–vintage Beaver. We were traveling in the same direction we had flown when we accompanied the oryxes to their home range, but now we continued on. For close to three hours the desert rolled beneath us, then we angled southwest toward the coastal mountains of Dhofar — and a seething mass of clouds. This is the only place in Arabia where the great Indian monsoon swirls ashore, engulfing the mountains in a drizzly gloom. The little Beaver plunged into the clouds; visibility dropped to zero. Somewhere below, unseen, the damp fog nurtured the trees that produced the finest grade of the world's finest incense: frankincense.

A half hour later, we dropped through a low ceiling and landed at the seaside town of Salalah. An hour or so after that, we were walking a beach. We would, we hoped, be walking back in time, into the land and life of an unknown people.

Our intent, on both this reconnaissance and the larger expedition we hoped would follow, was to sneak up on Ubar by first learning what we could about the frankincense trade and the People of 'Ad. Only then would we focus exclusively on the fabled lost city. There was good reason for this approach. If we *didn't* find Ubar, we could at least contribute something to the understanding of the region's history — and not completely disappoint our sponsors and the Omanis.

It was gloomy on the beach. In the season of the monsoon, the Arabian Sea was an expanse of dark, churning water. Palm trees shuddered and thrashed in the wind. We approached and entered the fallen gates of Sumhuram, a ruined city that had been partially excavated by Wendell Phillips's team in the early 1950s. The city was perched on a bluff overlooking a sheltered lagoon where, long ago, ships anchored to take on cargoes of frankincense. We spread out and explored Sumhuram's ramparts, dwellings, shops, storerooms, and a temple complex. We admired an elegantly chiseled inscription dating to the time of our People of 'Ad.

Inscription at Sumhuram

Years ago this inscription had been studied by Father Albert Jamme and the French epigrapher Jacqueline Pirenne. What we were looking at was a plaque commemorating the founding of Sumhuram, whose name is a composite word meaning either "the plan is great" or (more to the point) "the great scheme." Its six lines stated:

'Asadum Tal'an, son of Qawmum, servant of 'Il'ad Yalut, king
of Hadramaut, of the inhabitants of the town of Shabwa,
        undertook according to the plan the town of
Sumhuram, its siting and the leveling of the ground and its
        flow [of water] from
virgin soil to its putting in order. The creation and realization
        were on the initiative
and by the order of its master 'Abyata' Salhin, son of
        Damar'alay,
who is commander of the army of Hadramaut, in the country
        of Sakalan.

Nearby, three one-liners of scratched graffiti said: "The one-eyed [was here]," "Aywar and Hudail are dissatisfied," and "D E T E S T - A B L E !"[1] What was this about? A good guess would be that whoever hauled and chiseled the stones to build this place — namely Aywar and Hudail and a one-eyed chap — weren't all that happy to be here, which is understandable, as they were probably conscripted or even slave laborers.

Considering the formal inscription, it was evident that Sumhuram was *not* built by the People of 'Ad, the builders of Ubar. Rather, it was a colonial outpost of the Hadramaut, a kingdom whose capital "town of Shabwa" lay some five hundred miles to the west. Since Sumhuram was built "by the order of its master 'Abyata' Salhin . . . who is commander of the army of Hadramaut," its construction was likely a military operation, "a great scheme" designed to corner and control a lucrative sea trade in frankincense.

As we pondered this, the air was suddenly rent by a cry of "Kullu wahad fi haytan min shan aflan!" It was Ran Fiennes, proclaiming, in his best Arabic, "Time for a picture!" We gathered, and the camera took in a mixed, verging on motley, crew of several amateurs and a few professionals, none of whom had worked together before. And only Ran had been in this part of Arabia. Yet we had a shared enthusiasm, even as we posed for a group shot in a place whose stones said nothing of the existence of the People of 'Ad, or of their lost city of Ubar.

As the camera's self-timer buzzed, Ron Blom wondered aloud, "How, I wonder, do you say 'cheese' in Arabic?"

"Ghumda!" answered Jumma al-Mashayki, one of our Omani police escorts.

"Ghumda!" we all shouted, as the camera buzzed and blinked.

The gloom of the monsoon's overcast faded into darkness, but we lingered at Sumhuram. We were heartened to see that this had once been a splendid site, with finely finished masonry, clearly the work of an advanced civilization. We were disheartened, of course, to realize that its builders were not the People of 'Ad, but colonists from the kingdom of the Hadramaut.

By flashlight we took a last look at the inscription and its nose-thumbing graffiti. Angling the beam to cast the letters in deep relief, we picked out the name "'Il'ad Yalut, king of Hadramaut." His name dates the site's construction, for he is mentioned (as King Eleazus) in a Greek mariner's account written sometime between 40 and 70 A.D. Sumhuram, then, had to have been built no earlier than about 20 A.D.

It was then that it came home to us that we might truly be on to something. There are allusions to the frankincense trade dating back to thousands of years B.C., but Sumhuram had been built *after* the time of Christ. *Who, then, had managed the trade, shipped the region's precious incense over all those centuries before?*

Who other than our People of 'Ad?

For the next few days, we surveyed the coast with an eye to finding anything that might have been built by the 'Adites. We walked a couple of sites that might have been from their era, but they could also have been built by far-ranging Portuguese seafarers as late as the 1600s. Without actually digging, Juri explained, it was hard to tell. Depending on weather conditions and building materials, a site built within the last hundred years could look thousands of years old, and a thousand-year-old site could look as if it had been abandoned yesterday.

Wrapping up our survey and heading back to Salalah, we drove into a late afternoon patch of sunshine, a break in the pervasive gloom of the monsoon. Off to the left, Juri glimpsed something.

"Wait, wait! Over there!" he exclaimed.

At the wheel, Ran muttered, "Every time you see a rock, you want to stop."

"No! No. This is important!" insisted Juri.

Juri had spied an ancient graveyard, dozens and dozens of rock-walled mounds. Ran sighed and drove over to them; everybody got out and, led by Juri, prowled from one mound to another to another.

"Don't step on that," Juri cautioned Ran. "That's something right there. See that?"

He picked up a pottery shard and explained that it could have accompanied a burial and, over the millennia, worked its way to the surface. "Burnished ware. Look at that. See, hold it in the sun there. Kind of shines. See that? The people who made that pottery took a little stick and rubbed it real good to give it a shine. They couldn't make fancy pottery. But they tried hard. Did their best."

In a simple scrap of pottery, Juri the archaeologist had glimpsed the hand and life of an ancient potter. Moreover, the piece was unlike anything Juri had previously seen in Arabia. He logged the potsherd and hastened past the graves to the crest of a hill overlooking a marshy area called, we later learned, Khor Suli. He wasn't sure, but he thought he could discern traces of the docks of an ancient harbor. And closer to the sea we saw some structures that George Hedges dubbed "boats." They were stone enclosures, three to four meters long, shaped very much like small boats still in use on the Arabian coast. Juri wondered if cargoes of frankincense might have been sorted and weighed here before being loaded onto actual boats.

The site at Khor Suli almost certainly predated Sumhuram. Its masonry was rougher; there were no inscriptions. It had its own style of pottery, and its graves and stone "boats" were unique. This was the work not of outside colonists but of a native populace.

The People of 'Ad?

The next day we were to fly a long-range desert reconnaissance. If we were lucky, we would find compelling evidence of the People of 'Ad. Of course we might find absolutely nothing, in which case the quest for Ubar would probably be over.

In an early-morning drizzle, under a leaden overcast, we clambered aboard a camouflaged Huey helicopter provided by the SOAF, the Sultanate of Oman Air Force. It was a tight squeeze: our six team members plus three National Police escorts and the pilot and copilot. And camping gear, weaponry, water, and fuel.

Pilot Nick Clark, an Englishman on contract to the SOAF, flipped a sequence of switches. "Ignition," he announced. The Huey's main rotor turned, lazily at first. "Well, then, two minutes to liftoff." The Huey's rotors spun and whined, faster and faster. The big helicopter rocked and shuddered. And then, hardly realizing it, we were airborne, angling up into a thousand feet of dense monsoon. As we banked and turned north, there was a misty glimpse of the ground crew, waving and wishing us well.

A half hour later, we broke free of the coastal monsoon and saw before us the desert: blindingly bright, parched, pristine. Not a settlement, not a road to be seen. Nick the pilot swung an opaque combat visor down over his eyes. His voice crackled over the intercom: "Holding at 2,000 feet. On a direct bearing to your coordinates

18 degrees 32 minutes by 52 degrees 36 minutes. Should be there in a little over an hour." The coordinates were for the spot where Bertram Thomas, sixty years earlier, had crossed "the road to Ubar." Because we were heavily loaded and would be burning fuel at a rapid rate, we had elected to head directly to our most promising sites.

Nick: "Off to the left, that's the Wadi Ghadun." This great serpentine dry streambed heads north and into the sands of the Rub' al-Khali. What a trade route this could have been. I imagined caravans bearing frankincense off to the horizon, perhaps to Ubar. But imagining was one thing, and finding hard evidence was another. We could easily go down in the annals of Ubar exploration as hapless dreamers — "misguided at best."

I had read about cold sweats and seen a few in the movies. They hit Humphrey Bogart when he had his back to the wall and realized his automatic was in his other jacket. Wedged into the Huey, I remembered, or thought I remembered, a big close-up of John Garfield, as on a wing and a prayer he coaxed his battered B-24 back to Britain. I knew how he felt. This day would *have* to lead us somewhere, to something. I looked around. Was anybody else not feeling so good about this outing? Ran and George were lost in thought (conversation was impossible); framed by the barrels of a pair of automatic rifles, Kay smiled over at me, immensely enjoying her first helicopter ride. An old hand at flying desert terrain, Ron Blom shifted his gaze back and forth from the window to the Landsat 5 space image spread across his knees.

Nick the pilot, Ron, and I were linked by headsets. "Ron? Anything?" I asked him. "See anything?"

"Nothing as yet. Some great geology, of course. And up ahead it looks like we're in for a sandstorm."

"Afraid so," Nick confirmed.

I thought out loud, "It's hard imagining anyone actually *living* out here, isn't it? Now or then." Hoping that they would disagree.

"Yes, it is," said Ron.

"I'd say so," confirmed Nick, then added, "Coming up on target."

Discernible ahead was the ancient dry lakebed that had caught our attention on our very first radar space image. As we dropped down onto it, a cloud of swirling red sand, kicked up by our rotors, engulfed us. "Not to worry," Nick assured us, landing blind and hitting the desert floor. As the cloud of sand drifted clear of the idling helicopter, he warned, "Watch the rotors. Stay where you can see me. Nobody get behind me."

According to our calculations, we were more or less where Thomas had reported the hundred-yard-wide road to Ubar. But now we found the lakebed crisscrossed and churned up by modern vehicle tracks too narrow to show up on our space imaging. They would make it difficult to find and follow Thomas's road. How they got here was answered as three vehicles materialized on the horizon and sped our way.

An Omani military border patrol. Or, as we were to call them, the Phantoms of the Desert. They drove stripped-down, sand-swamped Land Rovers. No doors or windshields, but a few key accessories: racks for extra fuel and water, passenger-side .45-caliber machine guns, and, most critical of all, three extra batteries battened between the front seats. This wasn't the place for a balky starting motor. If you had to get out and walk, you might as well lie down and die.

We never saw the Phantoms' faces, hidden behind dark Afrika Korps–style goggles and woolen Omani head cloths called *shamags*. What they were patrolling for was a mystery to us, though we later heard they were engaged in a shoot-on-sight war with smugglers, who followed the route of the Ubar road as they ran drugs from the Arabian seacoast north across Oman and into Saudi Arabia. On their return, if they hadn't been gunned down, they would smuggle back gray-market color television sets.

The Phantoms of the Desert were quite willing to help us look for the road to Ubar. We thanked them for their offer, but considering the overlay of modern tracks, we felt we would be better off seeking our road farther on, in less trammeled reaches of the Rub' al-Khali.

Just as suddenly as they had appeared, the Land Rovers raced off across the desert to destinations and destiny unknown. And we were again airborne. Ahead now were the red dunes of the Rub' al-Khali. They did not yet form a solid mass of sand. Rather, they stretched in long rows, with intermittent gravel plains — known as "dune streets" — between them. At first these intervals were scored by modern tracks, but as we flew on — and the dunes rose to heights of two, three, four hundred feet — the tracks thinned out.

"Ten kilometers to target," Nick announced.

"Ron," I asked, "just to the north? You see what I see?"

"Could be our road. It's hard to tell with the blowing sand and all. But it's wider, more diffuse than the vehicle tracks we've seen."

"Older then?"

"Can't say." He squinted. "Can't see. Lost it."

The blowing sand blotted our vision, then cleared. And below us now there were no tracks at all. This, though, was as it should have been; below us, according to our best Landsat/SPOT image, a flash flood had wiped away our road.

Nick punctuated our thoughts: "Five kilometers to target." What we were heading for was the most dramatic appearance of the road to Ubar on our space imaging. It was *also* where, on his detailed map of Arabia, Bertram Thomas had spotted the *"Probable Location of the Ancient City of Ubar."*

"Four kilometers." Ahead now was a massive red dune more than six hundred feet high.

"Three kilometers." Nick guided the Huey over the shoulder of the dune and into a valley beyond.

"Two kilometers to target."

And there it was, unmarred by recent tracks: our road to Ubar. It came out from under the dune below us, continued for a good kilometer across an intervening plain, then disappeared under another dune. The track had to be very old, for it clearly had been laid down *before* the immense dunes that had buried it were formed. The road had been there for thousands of years.

"One kilometer . . ."

"Our road . . . Can you land right beside it?"

"Going in . . ."

We tumbled out of the helicopter and, as fast as we dared walk in the 115-degree heat, hastened across the plain. But the road wasn't there.

"You just walked right over it," Nick shouted (and no doubt chuckled to himself).

So we had. We marveled: the road to Ubar was clearly visible from 520 miles out in space, yet was barely discernible on the ground. What our space imaging had measured was not the earth's color or contrast, but its compression from the passage of untold caravans. With Nick's correction, we saw it: the road was composed of rows and rows of faint but unmistakable tracks heading northwest.

Quickly, we were airborne again for a short hop to the "L" site, which was the leading candidate for lost Ubar. Our expectations rose — and, even before we landed, fell. What might have been a walled settlement was nothing more than an unusual L-shaped alkali dry lake. It's been said that nature avoids right angles. Not here. The L formation had six of them, all beautifully shaped and geometrical, all formed by nature, not man.

As we lifted off, I thought again of the phrase "misguided at best."

Yet we saw that the Ubar road was still down there, and we were able to follow it onward, the sole track through this desert wilderness. Time and again, the ancient route would disappear into the sands, then, a kilometer or so later, reappear on an interval of gravel plain. Where it was buried, could Ubar also be buried? From the air, there was no way to tell. But on the ground, there would likely be telltale clues as the road neared the city: a concentration of potsherds, graffiti left on small rocks by camel drivers, perhaps even fragments of structures.

"I hate to say it, but if it's all the same to you, we should be turning back," said Nick. "It's hot. We're heavy. And we're not doing all that well on fuel."

Up ahead we could make out where the dunes became a solid mass, swallowing up the caravan tracks we'd followed. If Ubar was buried farther on, it would be impossible to find it. The helicopter banked, turned, and headed back along the road. We scanned it again, now looking for additional features that had caught our attention on our space images.

We passed back over the "L" site. A bust. Next we looked for what might have been a lost oasis where our space imagery showed a patch of infrared radiation. But whatever created it must have been transient — we saw nothing. (When the Landsat 5 satellite passed over, seasonal vegetation may have sprouted from a rain-dampened hollow in the dunes.) It was then that Nick announced, "We're not going to make it. We're burning fuel like mad." He hesitated, then suggested a plan. "Best bet is I drop everyone off, lighten up. Should be able to make it to an emergency fuel dump, then back. Okay?"

"Okay. But could you at least drop us off at the next waypoint, the one at 18 degrees 32 minutes by 52 degrees 31 minutes?"

"Will do."

This was a hot spot that had seemed promising on our SIR-B radar scan. But as we dropped down to it, we could see that it was yet another natural (though unusual) formation, a small limestone hill rising from the surrounding dunescape.

Nick offloaded us at the base of the hill. "Got to keep moving, so no shutting down. Watch yourselves. Should be back in an hour if all goes well . . ." In less than two minutes, he was on his way. His Huey became a speck on the horizon, then vanished.

We trudged to the top of the hill and found there a single, solemn bedouin grave. We checked the temperature. Shaded, at eye level, it was 120 degrees. The ground temperature, then, would be well over 160. We checked our supplies. Kay had some sandwiches. But most of our water, we realized, was still in the helicopter. We had only a few quarts in backpacks. Normally this would have been plenty, except that out here you were thirsty five minutes after your last drink.

Kay opened the umbrella she had brought along. Earlier on we had kidded her about it. Now, one by one, we took turns strolling over to Kay to double-check the time and gaze out across the desert, burning with a heat so intense it felt as if the oxygen were being drawn from the air. "Heat suffocation" sounded like an appropriate medical term — was there such a thing?

An hour went by. Everyone now just happened to be facing east-southeast, where Nick's helicopter had disappeared across the dunes. Our unspoken thoughts, I'm sure, were similar: what if Nick didn't quite make it back to the fuel dump? What if he ran out of gas or threw a rod or lost a bearing?

Two hours since Nick left. We said nothing, just listened. In the desert stillness, the tiniest of sounds was exaggerated. The crunch of a foot was thunderous, a whisper a shout. Then, two hours and twenty minutes after the helicopter had winged away, we heard a distant thumpety-thump, then spied far off a gnatlike, angellike speck skittering over the dunes.

Once we were aboard, Nick explained that the fuel dump had been marauded, either by drug runners or local bedouin. Out in the desert, fifty-five-gallon drums all but beg you to fill up. Luckily, he had managed to scrounge a few gallons and make it on to an SOAF outpost.

"Sorry about that, mates. It will be on to Shisur, then?"

"If it's okay by you . . ."

Prior to the drilling of recent bore holes, the well at Shisur had

provided the only reliable fresh water in this quadrant of the Rub'
al-Khali. As indicated on our space imaging, the Ubar caravan route
made a considerable swing to pass by Shisur. Our theory was that the
well had been a way station, a rest stop on the road to Ubar. It might
be a good place to find traces of the People of 'Ad.

A patch of green — a tiny oasis — marked Shisur. We circled once
and landed by a cluster of one- and two-room cinder-block buildings,
a seasonal settlement of the Bayt Musan, a band of the Rashidi
bedouin. They greeted us warmly, if a little warily, and offered us the
abiding hospitality of the desert. Little boys ran from house to house,
rounding up sufficient cups and glassware. Cardamom-flavored coffee was brewed and ceremoniously poured. Sitting in a circle on the
floor of Shisur's one-room schoolhouse, we inquired as to one another's well-being. It was a formal, almost courtly gathering. I sat on
my left hand to ensure that it wouldn't unwittingly reach for a handful of dates, a breach of bedouin manners.

The Rashidi were pleased we knew of their history. They listened
with interest to our idea and hopes of finding Ubar. Yes, they were
well aware of the lost city and believed it could lie as close as half a
day's drive away. Some day, Allah willing, a desert wind would bare its
walls. And did we know that there were ruins here at Shisur? Yes, we
had read of the fort here. It had been described by both Bertram
Thomas and Wilfred Thesiger. Thesiger noted that it had been built
by Badr ibn Tuwariq, a famed sheik of the early 1500s.

The Rashidi walked us over to Shisur's ruined fort. A lot of work
had gone into it, considering its location so far out in the desert. Too
bad that it dated back only five centuries.

"All right! Yes!" Juri had found a potsherd, burnished in much the
same way as the scrap he had found at Khor Suli on the coast.

"Unfamiliar," Juri mused. "Weird stuff."

"How old?"

"Could be the People of 'Ad. It's unique, could have been made very early on, a couple of thousand years ago. But it's also a little bit sloppy, see here? This rim. They didn't finish it as well as they could have. Maybe things weren't going so well, maybe they didn't care anymore. Could be late."

"Late . . . What do you mean by late?"

"Medieval, I suppose, or maybe even after that, at about the time Sheik What's-his-name built his fort here." The promise of the Shisur shard faded.

It had been a long day and was still a very hot one. As we returned to the helicopter, everyone dragged a bit, except pilot Nick, whose step was now remarkably sprightly. He said something about a need to conserve fuel, or maybe a need to avoid turbulence. The upshot was that the last leg of the flight was to be fast and low.

A half hour beyond Shisur, we banked and dropped down into the Wadi Andhur, a dry watercourse that originated in the incense groves of the Dhofar Mountains, off to the south. The wadi was once a major caravan route, a branch of the Ubar road.

As we followed the wadi south, I checked my watch: it was a little after seven P.M. The desert was no longer relentlessly bright and shadowless. The dry watercourse was cast in deep relief; boulders and patches of scrub brush were caught in the low sun's golden crosslight. They whipped by, no more than twenty or thirty feet below us. The wadi narrowed. We careened hard to the left, then to the right, then back again, following its twists and turns. Out the side window all was sky. A second later and the view was of the wadi floor, as the helicopter's rotors flattened vegetation and kicked up swirls of sand.

We hurtled on. Kay, by a window, was really enjoying the ride, untroubled by thoughts of imminent physical danger, such as catching a rotor and crashing.

Quite unexpectedly, the wadi widened. Ahead, in its center, rose twin mesas crowned with impressive ruins. We had come to the walled fortress of Andhur, reported by Bertram Thomas in 1930 and yet to be excavated or studied.

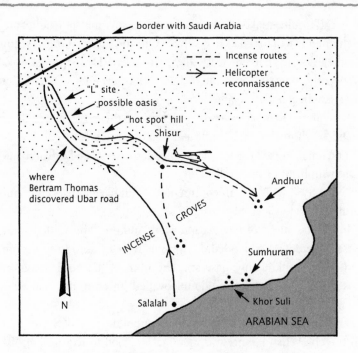

Reconnaissance into the Dhofar interior

In an aerobatic climax to the day, Nick spiraled into a deft landing within the walls of Andhur's south mesa. We offloaded our camping gear (and, this time, water). We improvised a plan: Nick would leave us to explore the site, then return, refueled, the next morning to pick us up.

We waved the helicopter on its way, then paused a minute to catch our breath and take in Andhur's splendid setting. It was quiet now, not the deathly silence of the open desert, but a stillness touched by a murmur of breeze, the chirps of a few hardy birds, and the bleating of goats.

A shout, in Arabic, came up from the base of our mesa. We looked over the edge to see a raggedy flock and a fierce Jebali (mountain man) herdsman. He was quite agitated. Ran listened.

"I can't quite make it all out, but he claims that our helicopter frightened five of his goats to death . . . Oh, and there's more. It seems another dozen goats, at least, have run away. Dear me, tsk, tsk." (When the occasion warranted, Ran did an excellent "tsk, tsk," at once skeptical and sympathetic.)

Ran offered to make amends. "Just bring us the frightened-to-death goats," he shouted down, "and we'll discuss a price."

The Jebali cursed (no translation needed), whacked his (surviving) goats with his staff, and stalked off.

"Nice try. Tsk, tsk," Ran commented. "You have to hand it to him for that."

In the last light of day, we prowled Andhur. The walled south mesa, where we had landed, had probably been used to sort, store, and guard frankincense harvests. The walls of the north mesa enclosed a curious half-buried double-walled structure that had likely been a temple.

Architecturally, the site was reminiscent of the port of Sumhuram on the coast. Its masonry was identical. It stood to reason that, like Sumhuram, Andhur was a colonial outpost of the kingdom of the Hadramaut. The site dated to perhaps 60 A.D., the work of wary outsiders (we thought) penetrating the land of our People of 'Ad.

We pitched camp and reflected on the long day's adventures. Though we had found and followed the Ubar road, we had made no startling discoveries. Three sites that could have been Ubar weren't. All told, we were pretty disappointed. Juri was intrigued by his pottery find, but, as Ran noted, you don't sell an expedition on a shard.

But we were not about to mope. The temperature had fallen, a full Arabian moon had risen, and ex-rocker George had brought his guitar. He now played for an audience of four adventurers, three Omani policemen and, possibly within earshot, a scattering of lost and traumatized Jebali goats. The ancient ruins of Andhur echoed with big-city blues and tales of Texas heartache, which somehow triggered a

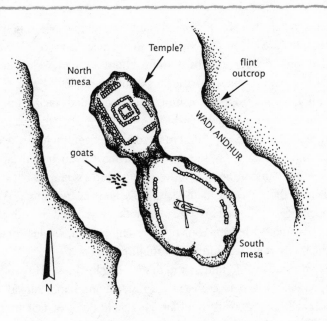

Ruins of Andhur

lively discussion of Junkyard Dog. Inexplicably, the American wres-
tler had become an Omani folk hero. Was it possible, our police
escorts wondered, that the great Dog might someday come to Arabia?

Civilizations rise and fall, come and go.

We turned in, all but the first of the policemen who would stand
watch in four-hour shifts throughout the night. The local Jebalis,
they had been warned, were not to be taken lightly; a few months
earlier they had murdered an outsider who had had the temerity to
venture into the Wadi Andhur and visit its mesa ruins.

As we stirred at dawn, our policemen reported that we had not been
alone in the night. They had seen lights and heard voices at the foot
of the mesa. With binoculars Ran swept the wadi. No one in sight.

After a quick breakfast of bread, jam, and coffee, we split up to explore the area. How close were we to the incense groves? Were the rock inscriptions reported by Bertram Thomas still in evidence? And what was that over there, on the far side of the wadi?

"Flint, an outcropping of flint!" Juri exclaimed, racing off. As I caught up, he explained that he knew of only one other source of flint in all Arabia. A deposit here would have been highly valued for tools and weapons. Andhur flint could have been traded far and wide.[2] "No wonder there's a fortress here," Juri exclaimed. "It stands to reason that . . ." He paused in midsentence, frowning. "What's happening?" I followed his gaze. Across the way, below the rim of the mesa, Humaid Khaleefa, our number-one policeman, lay sprawled in the rocks. Kay was running toward him.

We hurried over to find that in descending the mesa, Humaid had tripped, fallen, and severely gashed his arm and hand. Already Kay had broken open our first aid kit. He would be okay. Catching his breath, Humaid explained he'd been in a rush. But why? He gestured to the base of the mesa.

The matter of the frightened-to-death-and-runaway goats was upon us. The disconsolate Jebali herdsman of the day before had returned, backed by half a dozen of his clansmen, all armed with Belgian FAL automatic rifles. They had first encountered Ran. We joined him now and looked on as the Jebalis made their demands in heated Arabic. Listening to what they were saying, Juri whispered, "These are not nice people."

Ran turned to us and noted, in English, that we were outgunned and in a decidedly edgy situation. Nothing really scary, but we clearly had to make the right moves.

We had a couple of options. We could retreat to the high ground of the mesa and hold out until Nick's helicopter — due any minute now — returned. But that would be awkward and could even lead to an exchange of gunfire. Which would mean that we would never be able to return to ruined Andhur. Indeed, any sort of incident would mean never being welcomed back to Oman.

We decided that the better part of valor would be to negotiate an appropriate ransom. Our team and the Jebalis trooped to the top of the mesa, where we hastened to get whatever money we had from Kay, banker as well as paramedic. We had about $130 worth of Omani riyals, which we offered to the Jebalis. They scowled and shook their heads. Not enough.

They indicated that they might find items of value in our packs, so we invited them to look. Sure enough, one of the Jebalis was delighted by my plastic rain poncho. But his friends remained dissatisfied. We threw in a crate of apples. Still not enough.

Next they were into Ran's pack. They discovered and waved aloft a South Seas batik shirt. It was several sizes too large for any of them, but that was okay, it was enough. Our ransom was set. *Khalas* (deal done)!

Within a few minutes, our helicopter returned. As we ducked down to avoid the wash, Kay's notebook, open to get at our cash, snapped its binder and scattered our expedition records — and traveler's checks — to the reaches of the Wadi Andhur. Weeks later, a courteous yet understandably skeptical American Express agent allowed that losing-one's-checks-while-paying-a-ransom-for-goats-frightened-to-death was surely the most unusual explanation he had ever heard.

Returning to the monsoon-shrouded coast, we plotted a further helicopter foray out into the desert, this time to check out Ubar reports *not* associated with our space-imaged caravan route.

At this time of the year, many of the sheiks and elders from the interior could be found at the coast, relishing the cool and damp of the monsoon. Being born a son of the desert doesn't mean you have to love it. As one sheik wearily said, "The desert. Too hot. You know, just too much sun. A while ago, a German came by my tent. He told me of a place called Alaska. You know of this Alaska? *Schwea, schwea sams* (very little sun). *Wagan zain!* (wonderful)."

The sheiks of the desert lands of Mugshin, Mudhai, and Thumrait all had ideas about how and where we might look for Ubar. They spoke of a cave where, after the wicked city's demise, its treasure had been hidden. Surprisingly, they agreed as to its location, which they pinpointed on a map. There was also talk of an old man out in the sands who could take us to a stone signpost pointing the way to Ubar. There was, though, some disagreement as to the old man's health.

"He's dead," one sheik asserted.

"Not completely," another disputed. "He's strong, very, very strong. Couldn't be more than sixty percent dead."

"No, no. Eighty percent maybe."

Sort of dead or really dead, we were never to learn. That evening, pilot Nick contacted us with some bad news: the sandstorms we had encountered on our initial reconnaissance were now raging throughout the southern Rub' al-Khali. Even though his helicopter's engines were fitted with sand filters, it would be near-suicidal to venture north of the coastal mountains.

Two days later the sandstorms were undiminished in their fury, and reluctantly we headed back to Muscat and our flight home, thankful to have had at least a glimpse of the Rub' al-Khali and the road to Ubar.

Our police escorts saw us off. And one, Jumma al-Mashayki, had a last question for George Hedges. Throughout our reconnaissance George had relished consulting his pocket Arabic phrase book and constructing new and varied greetings, improvements on the usual "sabah al-heir" (good morning) or "khef halek?" (what's up?). Recently, though, George's greetings had elicited looks of distinct puzzlement.

Jumma was compelled to ask, "We policemen have all been wondering, Mr. George. Why, over the past few days, when you meet someone, why do you shake their hand and tell them, 'I am fire-wood'?"

Back in Los Angeles, Firewood, Ron Blom, Kay, and I reviewed the trip. It hadn't exactly been a disaster, yet it was well short of being a resounding success. The People of 'Ad and their lost city were proving to be surprisingly elusive. Though we had been reasonably frank about this with our sponsors in Oman, they didn't seem to mind, and looked forward to our return in the late fall. In the interval Ron Blom would have a chance to fill in gaps in our space imaging. In England, Ran Fiennes would plan and arrange the logistics of our next venture.

But then, five days after our return home, headlines announced: "IRAQ INVADES KUWAIT. SAUDI ARABIA THREAT-ENED." A war was on, a war that could well involve all of Arabia. The next day brought reports of Iraqi fighters dispatched to Yemen, where they were poised within striking distance of the Omani Air Force base that we had helicoptered into and out of the week before. We called and faxed our Omani contacts and friends to wish them well in the coming war. They thanked us and sincerely regretted that for the foreseeable future an expedition in search of Ubar would be out of the question.

12

The Edge of the Known World

WE PUSHED OUR PLANS BACK a year and hoped that when the time came we could still round up our original Ubar team. By winter the tide of the Gulf War had dramatically turned; Operation Desert Storm drove the Iraqis from Kuwait. In June 1991, Ran and George received word that we would be welcome to return to Oman, though parts of the Rub' al-Khali would be off limits. Also the Omanis reminded us that they expected the expedition to be filmed, which was fine by me, for that is what I do.

But who would pay for such a film? After our reconnaissance, we honestly had to admit that the odds were strongly against our finding Ubar. That was not the only problem. A Turner Entertainment executive had read our film proposal and decided "too much sand" (so that's what was wrong with *Lawrence of Arabia!*). National Geographic had already thought the expedition "too dangerous."

As our date of departure approached, George described the situation: "Everything is coming together, and we won't be able to shoot a foot of film." It was a Saturday morning, and George had stopped off to shoot a game or two of pool with his friend Miles Rosedale. Owner of the wholesale Rosedale Nursery, Miles was aware of our Ubar project and intrigued by the unique botanical properties of frankincense.

"A shame," Miles commented. "Five in the corner pocket." Clink-clump. As he eyed his next move, he asked, "Just what would a movie cost?"

A half hour later, George left with Miles's commitment to under-write equipment rental, eighty cans of 16-millimeter color negative, and the hiring of a cameraman and soundman.

George Ollen, a longtime free-spirited acquaintance, agreed to take a break from surfing and learn how to operate a Nagra location sound recorder. For the position of cinematographer we had an application written on stationery that featured a photograph of a man with a skull balanced on top of his head. The letter concluded:

Remember: I happily hung off the side of a balloon on a rope, rode 2000 miles through the deserts of Egypt and the Sinai on the back of a Harley, rode 1100 miles on a raft down the Yangtze River, A L L W I T H   A   C A M E R A   I N   M Y   H A N D . I'm bored with my day job and i'm ready to go. I even have a hat. Hope to hear from you soon.

Yours, Kevin O'Brien

In the second week of November 1991, all of our original Ubar team, plus cameraman Kevin O'Brien and soundman George Ollen, were back in Oman, at Salalah, on the shore of the Arabian Sea. After a fifteen-month delay, we were anxious to pick up where we had left off. Before we set out in search of Ubar itself, we intended to explore the Dhofar coast and mountains for tangible evidence of the People of 'Ad.

In 1329, Muhammad ibn Battuta, a traveler to the far reaches of the known world, visited these same shores and wrote that "half a day's journey east of Mansura [an old name for Salalah] is the abode of the 'Adites."[1] He was apparently referring to ruins encircling a great well that appears on Ptolemy's classic map of Arabia. There it is marked "Oraculum Dianum," the oracle of Diana, goddess of the hunt and the moon. (In reality, the site was probably dedicated to a southern Arabian equivalent of the Roman Diana.)

We had scouted this well during our first reconnaissance and found that it was called, even today, the Well of the Oracle of 'Ad.[2] Hidden in a valley just beyond the coastal plain, it was a very impressive hole in the ground, a good fifty feet wide and no telling how deep. Was it dug by man, or was it a striking oddity of nature? Whatever it was, we guessed that over the centuries it had gradually filled with debris, debris that could contain artifacts dating to the time of the People of 'Ad. Ancient peoples were forever dropping coins, curses (written on scraps of lead), offerings, and even vanquished enemies (and their weapons, armor, and all) into wells.

We backed a Land Rover Discovery as close as we dared to the rim of the well. Andy Dunsire, a stout, ruddy Scotsman, peered down it and muttered, "Daresay I doon like the look of that," then set about anchoring climbing ropes to the vehicle's rear bumper. Unique characters fetch up in the desert, and Andy was one of them, as were his associates Black Adder (Pete Eades) and Guru (Neal Barnes). They were engineers for Airwork, a British firm contracted to maintain Omani Air Force fighters, and one of our sponsors. As much as possible, Airwork would be giving Andy and his friends time off to help out with our expedition. Black Adder was fascinated by desert plants and flowers, Guru was a snake man, and Andy lived to explore desert caves and sinkholes, which made him just the fellow to lead the descent into the Well of the Oracle of 'Ad.

On parallel ropes, Andy Dunsire and ex–SAS commando Ran Fiennes roped up and, side by side, rappelled down the well's initial sixty-degree slope. "Ran, mind your . . . "Andy cautioned. "Uh, never mind," he added as a sizable chunk of rock broke lose from under Ran's foot, bounced down the well, and landed with a sickening *thunk* far below.

Juri hovered at the rim of the well. "Keep going," he shouted down.

"Easy for you to say."

"Ran, right to your left. That flat rock there. Can you check it out?"

Twenty feet down the well, Andy and Ran maneuvered to either

side of the rock Juri had pointed out and started clearing it with
trowels and brushes.

"Looks man-made," Ran shouted up.

Within a few minutes, Andy and Ran had cleared off four blocks,
part of what had once been a cut-stone platform — evidence that the
well was more than a natural hole in the ground. Gingerly now, Andy
and Ran eased over the platform and down into the well's narrowing
vertical shaft. And hardly were they out of sight than they reappeared,
pulling themselves up and out of the well.

"There's a bad overhang," Ran reported.

"And the walls. They're nothing but boulders stuck together with
mud," Andy added.

Dropping deeper into the well, they agreed, would be courting
disaster. The friction of the rappel ropes on the overhang could easily
trigger a rock fall.

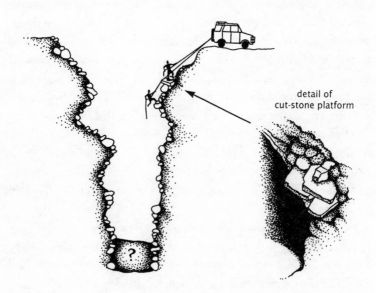

Well of the Oracle of 'Ad

For the rest of the day, Juri directed us and a dozen Airwork volunteers in surveying, clearing, and digging to the west of the well. He traced what appeared to be the outline of a small temple, perhaps a ritual entryway to the oracular well of the People of 'Ad. How were the oracles delivered, we wondered? By a *halmat*, a "seeress of dreams," or by *istqsam*, divination by the drawing of marked arrows? Or, Juri suggested, oracles could have been shouted up by a priest hidden out of sight in the well, possibly on the platform that Andy and Ran had cleared.

To find out anything more, we would have to devise a way to get down into the well. Ran turned to Kay and said ingratiatingly, "Tomorrow, Kay, I've been thinking. Maybe you could find us a crane? Say a construction crane? A big one? It would be ever so kind of you to do that."

Early the next morning, we were enjoying a cup of tea at the edge of the well when Kay cleared her throat with a distinctly self-satisfied "ahem." We looked over at her; she nodded out across the desert, where a great yellow crane was lumbering toward us. She had contacted British Petroleum, which had already promised us 8,000 gallons of fuel. So, she reasoned, it would be only a minor addition to their sponsorship if they lent us one of those big cranes they used to construct pipelines and oil wells and things like that.

The crane crept to the edge of the well, and Ran and I donned hardhats and clambered into its waist-high clamshell scoop. Out we swung, and down we went into the great hole in the ground until we could no longer see the crane's operator, or he us. The walls of the well closed in.

Ran called on a walkie-talkie: "Go left a touch, left a touch."

Ron Blom answered, controlled panic edging his voice, "Okay, we'll try. But I can't guarantee anything. He doesn't really speak English. The crane man doesn't speak English."

"Try gesturing," said Ran, then cautioned, "You see, if we touch the walls at all, we'll bring the whole lot down. So you've got to be very, very careful."

On down we went, past the platform Ran and Andy had cleared the day before. As Ran radioed instructions, I imagined Ron, out of sight far above, translating our instructions into signals, hoping he wasn't bashing us into oblivion. Did pointing this way and that mean the same in Arabic as it did in English?

"Uh, is this safe?" I wondered aloud, forty feet down the well.

Peering down, Ran didn't answer.

"Ran??"

"We'll see . . . we'll see," he replied, distracted by what lay below. "Strange place, this . . . Twelve feet to go . . . six . . . two . . . stop!" We jumped clear of the bucket.

The well was dry and would be ideal for excavation. Buried beneath our feet could be ancient offerings, clues to the identity of the People of 'Ad. Then we looked up. Overhead, tons of boulders were poised to break loose from the walls of the well. We whispered, as if the sound of our voices might bring them crashing down.

"These look really precarious, this lot . . . The bottom of that area there, if one goes, the whole lot will go." He shook his head, then looked to his feet. "And we're not alone, I see."

They had momentarily sought refuge in the well's debris when we jumped from the bucket. But now they were everywhere: scorpions darting in and out of tumbled brush, trash, and animal skulls and bones.

Ran and I agreed that our brief descent was treacherous enough; to try to excavate the well would verge on the suicidal. Climbing back into our bucket, we radioed Ron Blom to haul us back to the surface, where everyone was quite naturally disappointed by our report of conditions below. We had our crane, enthusiastic volunteers, and, as part of our gear, a ground-penetrating radar rig that could image the depths of the well and spot potential artifacts. We had even discussed running lights down so that we could dig around the clock in shifts. But now the only sensible thing was to pack up and move on.

Kay had set up a supply tent near the well, and now, as we broke camp, one of our volunteers hefted a last crate of supplies. And froze. "Under the box! S-snake!"

Kay looked over, let out a distinct "Eeek!" then turned pale. She realized that in moving things about in the tent, she had been — dozens of times — within striking range of the creature now coiled in the dust.

"Get Guru!" the cry went up from our Airwork volunteers. Guru was never without his curved snake stick, which he now used to coax the snake in the tent into a large bottle. "Nasty creature," he noted as he capped the bottle with a perforated lid. "Carpet viper. Hemotoxic. Neurotoxic. Hits you and you turn black."

"Oh, yes?" said Kay.

"No known antidote," Guru continued. "Hits you and you're dead in twenty minutes."

"Then it's nice," said Kay, "that he's in your jar just now."

"Yes, it is," concurred Guru, patting the sweat from his forehead. "Deadliest snake in the world."

Curiously, there was a lesson to be learned from Kay's carpet viper. Several classical authors had reported that the incense groves of southern Arabia were "guarded by flying serpents." The natural historian Diodorus of Sicily wrote that "in the most fragrant forests is a multitude of snakes, the color of which is dark red, their length a span, and their bites altogether incurable; they bite by leaping upon their victim." The historian Strabo added that they "sprang as high as the thigh, and their bite is incurable."[3] Despite such warnings by Diodorus and Strabo, the presence of any snakes in this land had long been disputed; it had been suggested that the "flying serpents" were in reality infestations of locusts. Or that they were apparitions concocted by ancient locals to warn outsiders to keep their distance. No, we learned, the "flying serpents" were really flying serpents. The mountains of Dhofar were full of them; on another occasion we saw one coil and strike. Though the creature didn't become airborne, it was fully capable of Strabo's "high as the thigh."

Classical Greek and Roman civilizations were well aware of Dhofar's coastal mountains, snakes and all, for they blocked the way to Arabia's fabled incense groves. These mountains were, in fact, almost certainly "Sephar, the eastern mountain range" that in Genesis 10:30 was taken to mark the edge of the known world. Throughout history, this escarpment had guarded a land stubbornly and persistently unknown, the heartland, we hoped, of our People of 'Ad.

While investigating the well of the Oracle of 'Ad, we had visitors, tribesmen who drifted down from the mountains. Their bearing was elegant; their hair, done up in fine braids and tinted blue, had the fragrance of frankincense. Members of the Shahra tribe, they spoke, in addition to Arabic, their own peculiar chirping, singsong language, called by early explorers "the language of birds."[4] They confirmed that, indeed, the well was still known as a well of the People of 'Ad . . . and one of their number, speaking in crisp, Cambridge-accented English, matter-of-factly told us, *You know, we are the People of 'Ad.*" His name was Ali Achmed Mahash al-Shahri, and where he would take us provided a major breakthough.

Like generations of Shahra before him, Ali Achmed was born and raised in the Dhofar Mountains. But as a young man he left his homeland to join the Trucial Oman Scouts, a military force that patrolled what was once a British protectorate in eastern Arabia. Commissioned as an officer, he was sent for further training to England. There he was surprised by the regard people had for traces and fragments of the distant past and by the great museums of London, Oxford, and Cambridge, built to house these artifacts. Ali Achmed realized that his people, the Shahra, had their own legacy: ancient writings hidden in the mountains of his childhood.

After mustering out of the Trucial Oman Scouts, Ali Achmed returned home. He sought out and hand-copied these pictographs, then acquired a Nikon 35-millimeter camera, taught himself how to use it, and went on to describe, document, and map dozens of sites.

Now he asked, could he show them to us?

With Ali Achmed as our guide, we drove up and over the steep sea-facing side of the Dhofar Mountains, emerging on a rolling, pastoral tableland. We passed Shahra herdsmen pasturing their small, short-horned cattle, the only cattle in all Arabia. Threading our way through a maze of tracks, we came to the mouth of a rugged and thankfully shady canyon. Even in December, the temperature was close to 100 degrees. On foot, we hiked on until we saw above us, on the canyon's south wall, a wide, shallow cave.

"You seek evidence of the People of 'Ad?" Ali Achmed asked. "Please, have a look."

After climbing up to the cave, we could see that its back wall was covered with hundreds of pictographs drawn in red and black pigment. Long ago, Ali Achmed explained, caravans had paused here and left their mark. He pointed out where someone had recorded the scene of a wolf attacking an ibex. Farther up on the cave's wall were three figures that looked like — but were not — the biblical Three Kings. Ali Achmed thought they were more likely three bandits, part of a group marauding an incense caravan.

Pictograph of wolf attacking ibex

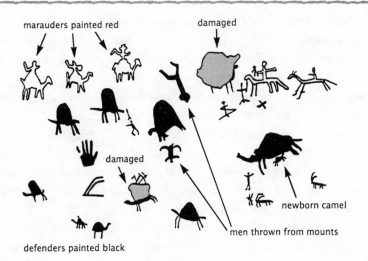

Pictograph of attack on a caravan

What excited Ali was that in the cave were not only pictographs but written inscriptions. Here was evidence that his ancestors not only traded in incense but were literate and civilized. (By definition, a civilization is a society with a written language.) Making sense of these meandering lines was problematic — a challenge beyond us.

Inscription in Dhofar cave

Many of the letters in these inscriptions related to letters in ESA, the Epigraphic South Arabic alphabet we had encountered on the coast at Sumhuram. But these mountain inscriptions contained eight *additional* letters, and it was anyone's guess how they were sounded or what they meant. Ali Achmed thought they might correspond to the eight sounds in the Shahra language beyond the twenty-eight sounds of classical Arabic. He pronounced them for us. They sounded something like *"sh'a, jh'a, je', k'e, ka', th', k'a, le'"* — curious, almost birdlike sounds, springing from the back of the throat. Sounds of Arabia long ago. Ali Achmed said, "Nobody can pronounce them but the people of our mountains."

At the far end of the cave, Ali pointed out a pictographic map that charted routes leading on to where, in antiquity, "silver" frankincense — the finest variety — grew wild and was harvested, as it still is. Ali Achmed showed us where.

A few days later we watched a band of little children dancing along behind two tribesmen — one wiry, one corpulent — as they crossed an arid valley and approached a scattering of scraggly trees with reddish bark. Bent and twisted, many of the trees were only waist high. Yet their resin, or sap, was once as valuable as gold. They were frankincense trees, found where the mountains of Dhofar gave way to the great interior desert of Arabia.[5]

The wiry man's craggy face was framed by a handsome white beard and a black turban. He wore a saronglike garment with a traditional silver dagger at his waist, complemented by a recent-issue assault rifle slung over his shoulder. Approaching a frankincense tree, he noisily exhaled, then chanted: "Ab st't d'h'la fe lh'ya!" (Exhale!) "Al as'r m'sly l'yo tr'le'ha!" (Exhale!) . . . His age-old song of harvest had a driving, intense rhythm, punctuated by strange, percussive exhalations.

Moving in time to his song, the wiry tribesman slashed bits of bark from the tree. A few yards away his partner — a *pasha*like fellow topped by a large red turban — mirrored his movements. The little children ran from one man to the other as, giggling and laughing, they played tag in groves of antiquity.

The Roman historian Pliny the Elder (23–79 A.D.) relates that these frankincense groves were inaccessible and that the southern Arabians — our People of 'Ad — did not encourage visitors. In addition to tales of flying serpents (quite true), the 'Ad appear to have propagated stories of deadly vapors arising from the punctured trees (not so true). Not surprisingly, Pliny tells us that "no Latin writer so far as I know has described the appearance of this tree." Nevertheless, he learned that "the district . . . is rendered inaccessible by rocks on every side, while it is bounded on the right by the sea, from which it is shut out by cliffs of tremendous height. The forests extend twenty *schoeni* [about five miles] in length and half that distance in breadth . . . In this district some fifty hills take their rise, and the trees, which spring up spontaneously, run downward along the declivities to the plains."

That was as we found it: the finest trees, those that produced silver frankincense, were confined to a surprisingly small area and were wild. They stubbornly resisted cultivation. Pliny further informs us:

It is the people who originated the trade, and no other people among the Arabians, that behold the incense-tree; and, indeed, not

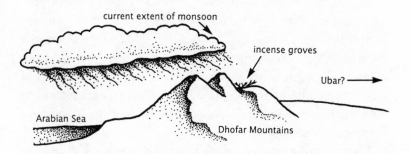

Cross-section of Dhofar Mountains

all of them, for it is said that there are not more than three thousand families which have a right to claim that privilege, by virtue of hereditary succession; and that for this reason those persons are called sacred, and are not allowed, while pruning the trees or gathering the harvest, to receive any pollution, either by intercourse with women, or coming in contact with the dead; in this way the price of the commodity is increased owing to the scruples of religion.[6]

This holy harvest of frankincense, some scholars feel, was a convenient fiction, an invention of southern Arabians intent on hoodwinking credulous customers. Yet we were inclined to believe Pliny's account. There was something very serious, almost formalized, in the way our two tribesmen moved from tree to tree to the rhythm of a measured chant.

The chant ended with a loud exhalation. The tribesmen and the children drifted off across their land, a moonscape dotted by small groves of frankincense. The shouts and distant laughter of the children dissolved into a desert breeze, which now bore the piny, slightly raw scent of freshly cut frankincense. Each slash in a tree's bark produced a dozen or so thick white globules of resin. Slowly these globules would lose their milky opacity and gain a silvery translucence as the frankincense hardened and crystallized. Fifteen days hence, the men would return to scrape it into special shallow baskets. Though a portion of the harvest would be kept for their own use, most of it would be traded to the coast. It would be used to sweeten the air of households throughout Arabia, to scent men's beards before dining, to fumigate robes and dresses. Little kids would chew it as gum. It would be a prized ingredient in exotic perfumes, including the French-Omani scent Amouage, promoted as the most expensive fragrance in the world.

That we might see more of the living history of the highlands of Dhofar, Ali Achmed invited us to visit a remote Shahra settlement.

Driving by night, we arrived at dawn at a compound of four thatched huts clustered around a brushwood corral. Three of the huts sheltered cattle; the fourth was home to an extended family. Though the hut was windowless, two doors let in sufficient light to illuminate the single large room. Its walls and domed ceiling were woven of twisted, blackened tree trunks and branches, the best wood to be had in an arid land. Two young girls were rolling up sleeping mats. A baby was squalling in the corner. Two older men and a woman crouched by an open fire, making their preparations for the day, a day measured by the burning of frankincense.

Though the woman wore a long, hooded black dress, she was unveiled. A gold ring pierced her nose, her eyes shone with self-assurance. She was the settlement's matriarch. With brass tongs she picked embers from the fire and placed them in a brightly painted clay incense burner shaped like a horned altar. Then she added crystals of frankincense, which glowed brightly and immediately gave rise to a fragrant, smoky plume. All the while she chattered with the two men in the Shahra's strange "language of birds."

"Incense is most pleasing to God," she said, adding more crystals.

"But enough, woman, enough!" interjected one of the men, his eyes smarting from the smoke.

"Too bad for you," she said, laughing, and led the way outside. The men downed handfuls of pine nuts, the last of their breakfast, and followed.

With clouds of incense billowing skyward, the little group circled the compound's corral. And in the light of day we saw that the men were wearing elegant purple robes looped over their right shoulders, a rarely seen traditional dress. They paused to offer prayers and incense at the entrances to the three domed huts in which their cattle had spent the night. The incense wasn't to offset the smell of the cattle (though it helped); rather, it was offered to protect the animals — from djinns.

The mountains of Dhofar, the Shahra believed, were rife with djinns, invisible spirits born of smokeless fire. By day they dwelt at waterholes and in dark gullies. Though some djinns were friendly, most were not. Given to inflicting misery and misfortune, they could take the form of whirlwinds and raging sandstorms. Or they could shape-change into reptiles, various beasts, or even humans. Their true identity was discernible only by their feet, which were like the hoofs of asses. In great numbers, djinns were abroad at night, especially on Wednesdays and Fridays. Flying out across the land, they uttered screams so loud and penetrating that anyone unwisely out and about would lose his wits. It was a time to bar doors and windows and leave the darkness to its owners.

Now, in the early morning, frankincense dispelled any lingering djinns, and the cattle could be led off to pasture with a reasonable assurance of safety. The herdsmen would nevertheless be wary of strangers going their way. In broad daylight, djinns could manifest themselves as fellow travelers, leading men and animals astray, often to their deaths.

The cattle disappeared over a hill, their passage marked by a lingering cloud of dust that hung motionless. "A good sign," Ali Achmed noted, looking off across the land. "No djinns."

The Shahra remaining at the settlement turned to their daily tasks. The dwellings of men and beasts were swept out. A little girl worked at a loom; her mother churned milk in a leather bag. A dog lazed in the sun, one eye open, lest it be kicked by a nearby goat.

Despite everyone's diligence, it appeared the djinns had worked some mischief. A little boy hadn't been able to shake off a bad cold, and something needed to be done. The settlement's matriarch added fresh frankincense to a burner and led the child to the center of the corral. Round and round she circled him, enveloping him in incense. She chanted, "Look at this your sacrifice: frankincense and fire. From the eye of the evil spirit; of mankind, from afar; of kindred, nearby, and from afar. Be redeemed from the evil spirit. Look at this your sacrifice: frankincense and fire."[7]

She passed the burner to an older man, who continued the ritual, circling and chanting. Frankincense and fire were a potent combination. The incense brought the blessing of Allah, and fire — even a small spark — was believed even more effective than the name of Allah in curing possession. Fire dispelled djinns, creatures born of fire.

As the day wore on, tribespeople came and went. A woman came by in search of a lost chicken. Assault rifles jauntily cradled in the crooks of their arms, three young men stopped by for coffee; they had been down to the coast and shared news of the outside world.

With Ali Achmed translating, we asked about the People of 'Ad. Yes, they all agreed, the 'Ad were their long-ago ancestors. The Shahra knew about Ubar and referred to the city's inhabitants as "Irema." We were startled by this, for Irem (or Iram), we believed, was the Koran's name for Ubar. Here was a living link between the two principal names for our city![8] The Irema, the Shahra told us, were a rich if wicked people who ate off golden plates. That is, until "their city turned over."

A young man with the mustache of a brigand chuckled and told us that to the Shahra the phrase "Take him to the Irema" means "Get rid of him."

The oldest of the group regaled us with the tale of how the golden treasure of Ubar, spirited away from the doomed city, was to be found in a desert cave, guarded by a snake.

"Was the snake a djinn?" I asked.

"How could that be?" he shot back. "Do snakes have feet?"

I realized my mistake. To be a djinn you have to have hoofs, so a snake can't be a djinn. We agreed that snakes can still be pretty nasty.

The storyteller continued, "One way to keep the snake away [from the treasure] was to have a holy man read to it." This, he said, led two thieves to find themselves a holy man and have him read to the snake while they helped themselves to Ubar's riches.

The old man's voice fell to a whisper: "You see, they were going to cut the holy man out. But he, being both holy and wise, suspected this. He stopped reading. The snake ate one thief; the other ran away. Nobody's gone there since."

The thought crossed my mind: if we couldn't find Ubar, maybe we could find the cave. Read to the snake.

Late in the afternoon, the family's cattle reappeared, and with whistles and trilling shouts, were herded back into their huts. In the clan's living quarters, incense was again burned, and a *majlis*, a communal gathering, began. In their ancient language, the Shahra sang a rousing song of revenge, then shifted to a melancholy melodic recital of lost love and found wisdom. They sang a capella and antiphonally, with evenly divided groups of men answering each other. They drew their voices across the words like bows across strings, as if to echo a psaltery of the past. In this isolated settlement, speech and song — and a way of life — had been preserved since the ancient days of the incense trade.

At the end of a long and rewarding day, we took leave of the Shahra and headed back to Salalah. We kept an eye out for djinn, but, it being Thursday, we were unmolested. (It's Wednesdays and Fridays that you must beware.) As we bumped along the dirt track, we passed other Shahra settlements, illuminated by kitchen fires, occasional gas lanterns, and the passing glare of our headlights. If the Shahra live now as the People of 'Ad once lived, it was understandable why tangible remains of the 'Ad had proved so elusive. Their culture may have been complex and literate, but their surviving artifacts would have been few. The impressively domed dwellings and most of their contents were perishable. About all that would survive for even a century would be fragments of fired pottery, foundation stones, and the stones marking their departure from their life and land.

Their graves.

13

The Vale of Remembrance

THE VERY NEXT DAY, Ali Achmed took us beyond the settlements of the Shahra to a long, meandering valley — the course of the Wadi Dhikur — that dropped from the tableland of the Dhofar Mountains to the desert beyond.

The Shahra knew the valley of the Wadi Dhikur as the Vale of Remembrance, for this was where they had laid their dead to rest for hundreds, even thousands, of years. An eerie mist shrouded the valley's upper reaches, draining the color from its rock walls. The air was clammy, and not a breath of wind stirred. The place was oppressive. The first graves we encountered were Islamic, oriented to Mecca. Two stone slabs marked a man's burial, and three marked a woman's. Ali motioned to where, higher on the valley's wall, stone blocks sealed a series of caves. We scrambled up and peered into a breach in the stonework. From the darkness within, a congregation of skulls returned our gaze. Patches of their head cloths were still intact. Other bones were scattered about the cave. Since it was not oriented to Mecca, this tomb appeared to be pre-Islamic, earlier than the 600s. We could have reached in and collected sample textiles for carbon dating, but we didn't, for we felt ourselves outsiders to this majlis of the dead. They, not we, belonged in the valley.

Returning to the valley floor, we shared, rather quietly, the lunch Kay had laid out on the tailgate of a Discovery. Ali Achmed ate only half of his turkey sandwich; the rest he scattered in the air.

"Ali? Looking after the birds?"

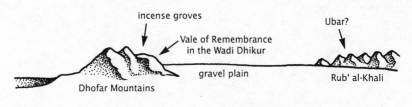

Cross-section of the Dhofar region of Oman

"Feeding the djinns," he replied with a laugh that wasn't really a laugh. Even to the enlightened, djinns are not to be dismissed. They stalk the pages of the Koran, where they are considered, after man and animal, a third creation of Allah. 'Afrits, a particularly troublesome brand of djinn, dwell in graveyards. They are the spirits of wicked men turned away from Paradise, doomed to haunt their place of burial.

"And who was wickeder than the People of 'Ad?" asked Ali Achmed, paraphrasing a verse from the Koran.

We drove on down the Vale of Remembrance. Shading his eyes from the afternoon sun, Ali scanned the widening valley.

"The stones," he said, "you must see the stones. They're part of this."

He directed us to a row of stone monuments, each consisting of three unfinished slabs of stone tilted together to form a crude two- to three-foot-high pyramid, known as a trilith ("three-rocks").

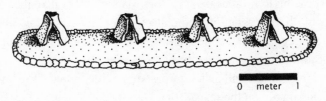

Dhofar triliths

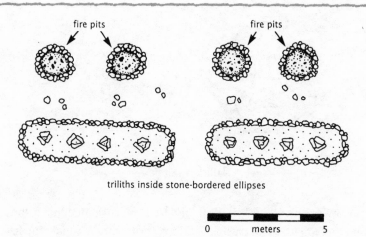

fire pits   fire pits

triliths inside stone-bordered ellipses

0   meters   5

Line of triliths viewed from above

Rows of triliths have been reported throughout southeastern Arabia. This site had eight pyramids in a line and, parallel to them, four fire circles. Elsewhere we found rows of as few as three and as many as twenty-five triliths.

The triliths of the Vale of Remembrance were as old as the People of 'Ad. Ali Achmed showed us where, the year before, he had dug beneath two of them and discovered ash, and — via an English friend — sent it off for carbon dating. His ash samples dated to 60 and 110 B.C. (plus or minus approximately one hundred years).

Ali was convinced that the triliths had once been "quadriliths," that each set of three stones once supported a capstone, which over the years could have easily tumbled off. Spotting a rock that to us looked like any other, he crouched down, dug around it, and turned it over. It was inscribed with well-preserved letters from the same alphabet used in the rock art of the mountain caves.

His fellow Shahra, Ali Achmed told us, held that triliths were memorials to the ancient dead, a reasonable idea considering the

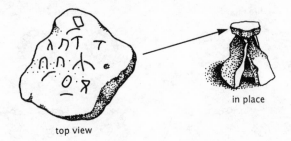

in place

top view

Trilith capstone

funerary nature of this valley.[1] It has also been suggested that triliths were route markers for passing caravans or sites for the ritual crucifixion of ibexes, desert animals whose crescent horns symbolized the crescent moon, the principal deity of pre-Islamic Arabia.

That afternoon we visited several trilith sites and saw many others in the distance. They were generally on flat terraces above wadi drainage channels and were often silhouetted against the sky. Why were they here? What were they for?

Sometime later I came across a tantalizing passage in Hisham ibn al-Kalbi's *Book of Idols*, a catalogue of gods of pre-Islamic Arabia compiled in the early 800s. Al-Kalbi recounts, "Whenever a traveler stopped at a place or station [in order to rest or spend the night] he would select for himself four stones, pick out the finest among them and adopt it as his god, and use the remaining three as supports for his cooking pots. On his departure he would leave them behind and would do the same on his other stops."[2]

Here was a validation of Ali Achmed's opinion that these monuments had been constructed of not three but four stones. And the fourth, the missing capstone, would have served as a traveler's god, a *betyl*. The word expresses a concept common to both the Old Testament and the Koran. It comes from the Arabic *bayt-el*: the dwelling place *(bayt)* of God *(el* being the root of the word *Allah)*. In Hebrew, *bethel* has the same meaning.

A betyl was a rock where God dwelled, or at least visited. Or several rocks where several gods took up residence. In Semitic tradition it was important that the rock not be representational, that it be without countenance. Arabs and Jews alike have long abhorred the idea of graven images. Idolatry to them was a crude invention of the Mesopotamians. If God (or gods) descended from the heavens and was to be found on earth, it was as a spiritual force — *Sakina* in Arabic, *Shekinah* in Hebrew — that dwelt in a featureless rock. A betyl.

A goodly number of the "standing stones" cited in the Old Testament were betyls. And, a legacy of ancient times, the betyl of all betyls is the Ka'aba, the windowless black basalt cube at the heart of Mecca. It is revered as the Bayt Allah (same as *bet-yl*): the dwelling place of God. An Islamic tradition has it that the very first Ka'aba was constructed in heaven, where it remains to this day. After his expulsion from Paradise, Adam built the first earthly Ka'aba on a spot directly beneath the heavenly one. God was pleased and appointed ten thousand angels to keep watch over the site. But these angels, despite their numbers, were remiss in their duties. Adam's Ka'aba fell into disrepair and was destroyed by the Flood. After Abraham traveled to Arabia from Syria, it fell to him to rebuild the structure with the assistance of his son Ishmael, who became the progenitor of all Arabs.

Whatever the veracity of this primordial history of the Ka'aba, it is certain that long before the advent of Islam, pilgrims from all Arabia journeyed to Mecca. They worshipped the holy cube by circling it, by anointing it, by touching and kissing it. As the mirror of a heavenly archetype, it had an elemental appeal. According to the late Iranian philosopher Ali Shari'ati, the glory of the Ka'aba was that it was no more or less than a simple cube, representing "the secret of God in the universe: God is shapeless, colorless, without simularity. Whatever form or condition mankind selects, sees or imagines, it is not God."[3] Though the pre-Islamic Arabs may have been reaching out to more gods than one, this was the essence of what drew them to Mecca — and to the veneration of betyls throughout all Arabia.

There were once rival Ka'abas in the Arabian cities of Nejran and Sana'a, and in remoter areas stones of all sizes and shapes were worshipped. George Sale, an acerbic Arabist of the 1700s, noted that pre-Islamic Arabs "went so far as to pay divine worship to any fine stone they met with."[4]

To me, the triliths of Dhofar — with their inscribed capstones — appeared a natural and logical part of all this. They were betyls, each inhabited by a god's essence. But also, on a homely level, as al-Kalbi noted in his *Book of Idols*, triliths supported the traveler's cooking pots — they were his stove as well as his shrine. This juxtaposition of the sacred and profane might at first seem jarring, but for a practice that has been documented at Petra, a northern terminus for Arabia's incense trade routes. Betyls large and small were everywhere at Petra, and their worship was an everyday event. After worship, groups of thirteen men would often ritually assemble inside rock-cut family tombs to dine in the presence of their departed ancestors. In their honor, the men would eat off golden plates and drink wine from a shared golden goblet.

With clay bowls and cups instead of golden ware, similar rites could have taken place twelve hundred miles to the south, on the far side of the Dhofar Mountains. It was not hard to imagine the People of 'Ad ritually circling the triliths and anointing their holy capstones with water, oil, or even blood. The capstones could then have been removed, and a ritual meal prepared in honor of the tribe's departed ancestors, who were all about, entombed in the valley walls and likely underfoot as well.

Here the People of 'Ad would have sought the blessings of their ancestors and gods before trekking out across the vast desert to the north. On their way to their mysterious city. The city we sought: Ubar.

14

The Empty Quarter

JUST BEYOND THE DHOFAR MOUNTAINS lay Thumrait, an Omani airbase that would serve as our staging area in the search for Ubar. Airwork, the British company handling aircraft maintenance there, kindly offered us sleeping quarters, good food, and a place to store everything from an 8,000-gallon gasoline tanker to frozen food to boxes of dental picks (for touchy bits of excavation). At Thumrait we sorted our desert gear, two truckloads of it driven from Muscat. With the aid of Airwork volunteers, Ran set up a thirty-foot antenna for long-range communication.

We took stock of what we had found. On and near the coast, we had been hard put to find evidence of the People of 'Ad, but in the mountains, Ali Achmed had shown us their cave art, the work of an imaginative, literate people. And we had seen that the Shahra continued their incense-oriented way of life to this day. In both mountains and desert, we had puzzled over rows of triliths, monuments that Juris Zarins felt might mark the homeland of the People of 'Ad.

Beyond Thumrait, though, there were no more triliths. And, for all we knew, no more of the People of 'Ad, except for fragments of a road, which we hoped led to Ubar.

With no resupply possible beyond Thumrait, our plan was to strike out for the dunes to continue the search where our 1990 reconnaissance had left off. Ran and Kay worked out how much fuel, water, and food our Land Rovers would carry and estimated that we could be self-sustaining for five days. If nothing much went wrong, we could get to the border of Saudi Arabia and back. If we found anything of consequence, we would return to Thumrait, load up, and head back out.

Friday, December 13. Day 1: to the dunes. After a tasty breakfast of pancakes and fresh fruit ("the Last Breakfast," punster Juri called it), we packed our three Discoverys. Our team was joined by desert-knowledgeable Andy Dunsire and a lean, bespectacled fellow bearing a frying pan, two pots, and a battered suitcase.

"Mr. Gomez, our cook," said Kay, introducing him all around. "On loan from Airwork."

"I come here as a guest worker from the country of Goa. In Goa, all cooks named Gomez," he said, or at least that's what we thought he said, for he spoke very fast, as they apparently do in Goa. ("English on speed," cameraman Kevin O'Brien called it.) When Mr. Gomez accelerated beyond a sentence or two, which he now did, apparently listing some of his favorite dishes, it was ultimately only Kay who could understand him, and she admitted to catching only every third word or so.

As Mr. Gomez elaborated on curry in its various forms (or so we thought), someone wondered if we really needed a cook just now, but backed off when Kay asked, "You want to deal with these things?"

"These things" were fifteen sturdy cardboard boxes of Meals-Ready-to-Eat (MREs) left over from the Gulf War. Though they had overrun their expiration dates, we had been told their contents would last forever. "Curry sounds wonderful, but I'm afraid we'll have to make do with these for now," she explained to Mr. Gomez, then assured him, "You'll have real food to cook later on."

"Out in the Rub' al-Khali, will Mr. Gomez be comfortable in what he's wearing?" ventured Ran diplomatically, noting that at the moment Mr. Gomez was dressed in cook's whites and Chinese slippers.

"Most certainly, yes, why not?" Mr. Gomez was quick to answer.

It did seem a bit unusual taking a uniformed cook out into a place described by the normally calm *Cambridge History of Islam* as "the most savage part" of Arabia, "a veritable hell on earth." But in Mr. Gomez's words, "Yes, why not?"

The MREs were lashed to the roof rack of a Discovery, joining sand ladders, sleeping bags, and fifteen jerry cans of water and fuel per vehicle. Plus clothing, blankets, camera equipment, automatic rifles, and so on. We had an alarming amount of gear.

"It's a mystery," Kay said of our three Discoverys, shaking her head, "why they don't just sink down to China."

"Kansas," corrected Ron Blom. "From here in Arabia, they'd sink down to Kansas."

We shoehorned ourselves in among our gear and were on our way. Guided by Andy Dunsire, we picked up an old track that headed north across a vast, flinty, featureless plain. Our stout Discoverys, to our considerable relief, shouldered their loads well, churning on through deep ruts, then loose sand, then absolutely miserable ruts, and finally a mix of fine sand and gravel that allowed us to pick up speed and cruise at a surprisingly good clip, a steady forty miles an hour.

The landscape changed. A dreary plain became a plain of illusions. Off to the left was a pale blue, shimmering lake. Soon there were lakes to the right and all about. The Discovery ahead of us, driven by Ran with Andy at his side, splashed into one. We followed and never got wet.

The vehicles ahead of and behind us became rippling abstractions, mirrored upside down in water that wasn't there. We were floating out across Arabia, our destiny uncertain. It didn't at first register when Ran's vehicle fishtailed wildly and barely recovered. Andy radioed back, "Watch it! Camel wallow!"

The wallow — a place where camels come to roll on their backs and take sand baths — was hidden by a mirage. They hadn't seen it coming. We braked and hit a twenty-foot patch of treacherously loose, deep sand, almost quicksand. How we got through it without spinning out of control or bogging down hopelessly, I'm not sure. Talking by radio with Andy, we then learned that the trick to negotiating a camel wallow is to slow down, engage the vehicle's differential lock, and don't try anything fancy with the steering wheel. By late afternoon we had white-knuckled our way through three more wallows.

An unexpected weather front moved in, clouding the sky and dramatically dropping the temperature. The desert's phantom lakes evaporated. Toward evening we sighted on the horizon the first dunes of the Rub' al-Khali, in English the "Empty Quarter."

The Empty Quarter derives its name from a legend that on the eve of creation God divided the world into four quarters. One was the sea, two were set aside as the settled lands, and the fourth was to be forever barren: the Empty Quarter. Sprawling across a quarter of a million square miles of the Arabian interior, it is a desert of dunes, a vast sand sea, the largest on earth. In 1885 a Colonel S. B. Miles was one of the first westerners to gaze upon it from our direction, the south. He had this to say: "This wilderness . . . stretches away to the westward for about 700 miles, forming the largest and most inhospitable expanse of sandy waste on the continent of Asia. Broadly speaking, it is devoid of rivers, trees, mountains, and human habitations, unexplored and unexplorable, foodless, waterless, roadless, and shadeless, windswept, and a land of quietude, lethargy, and monotony, perhaps unparalleled in the world."[1]

Over the years, the Rub' al-Khali's reputation has gotten no better. There is something inherently terrifying about so much sand and so little life. In the 1930s, Bertram Thomas saw it as a place of romance and wonder but also as "a hungry void and an abode of death."

Our personal reflections on the Rub' al-Khali would have to wait. As the sun edged to the horizon, we pulled up in the lee of a large dune and hastened to set up camp. Kay and Mr. Gomez sorted our rations of MREs. There were twelve assorted meals to a box, which had to be matched to the dietary requirements of our twelve-member team. Our Omani Police guards were yes on beef and no on pork. Our camera crew were vegetarian, so they got (the closest we could come) the chicken and turkey dishes. By a process of elimination, Kay and I wound up with — breakfast, lunch, and dinner — a choice of sliced ham or franks and beans ("NOT FOR PREFLIGHT USE," the label advised). The MREs weren't bad. They could be eaten cold, were better boiled, and were best fried; for added zing, their packets included tiny bottles of Tabasco sauce.

Our sheltering dune was clearly visible on our space imagery, providing Ron with a good opportunity to verify the accuracy of our satellite navigation system. He punched out instructions on the receiver's keypad, only to have it flash "NO SATS FOUND." How could this be? By his calculation, even in our remote location three satellites should be overhead. Ron shut the receiver off, then tried again. And again. The device was adamant. "NO SATS FOUND."

This was serious. Satellite navigation had proved so reliable on our reconnaissance that we had planned to find our way solely with satellite fixes plotted directly onto space images. What was wrong? Ron guessed that the satellites were up there, all right, but had been made inaccessible to civilian codes. "Could be the Gulf War has fired up again, and we want to cut off enemy access to precise navigation. Beats me."[2]

On that uneasy note, as darkness overtook us, we each spread our allotted two blankets on folding aluminum camp beds. These put us a few inches above any creatures that might go slithering or skittering through the night. As we bedded down, Ran propped himself up on his elbow and casually mentioned that we were in the domain of a solpugid commonly known as the camel spider.

"Not asleep yet?" he queried, to be sure no one missed his tale. He then recalled that some eighteen years earlier, he and his Omani military patrol had also camped on the edge of these dunes.

"My signaler, Ibrahim, got visited in the night. The spiders are six inches long, hairy legs and big mandibles."

"For their size, I daresay," Andy Dunsire chimed in, "they have the strongest jaws of any creature in all the animal kingdom."

Ran continued, "One of them couldn't get into Ibrahim's sleeping bag, so it started to eat his face! It desensitizes before it bites, so that you don't know it's biting. This fellow woke up in the morning, and half his nose and all of his cheek had gone AWOL!" Ran let this last thought hang in the desert air, then signed off with a cheery "Sleep well!"

Saturday, December 14. Day 2: into the Rub' al-Khali. The weather front that had moved in the previous afternoon had driven the overnight temperature down to near freezing. Some of us had stoically made do with our two blankets, others had tried to sleep sitting up in the Discoverys. A rather droopy Mr. Gomez, a gray blanket draped over the shoulders of his cook's whites, brewed an inky pot of coffee and doled out a round of MREs. The combination perked us up. "MREs, we love 'em," Kay commented. "All protein and sugar. Fighting food."

We packed up and set out northwest across the dunes. Our first objective was the Wadi Mitan at a point where it terminated in a distinctive dry lakebed, twenty kilometers long. With our satellite navigation inoperative, this would be a reliable waypoint.

For what lay ahead, there were two theories — two extremes — of how to handle your vehicle. The first, widely practiced by the bedouin, was a kind of Zen of dune driving. They would "read the sands," evaluating slope, texture and color to determine exactly what path to take at what speed. They would then effortlessly float out across the terrain. No sideslips, no spinning wheels. It looked so easy, so effortless. To the skilled and confident bedouin, it was.

The other way was our way. Not having the faintest idea if the sand ahead was hard packed or treacherously soft, we careened recklessly up and down, over and around the dunes, foot to the floor, driving as fast as we dared. It looked like joy riding. It *was* joy riding. But it was also a survival tactic. To slow down was to risk getting stuck.

With Ran at the wheel, the first of our Discoverys crested the shoulder of a dune and for a second or two flew through the air. "Whaa! Ha ha!" Ron shouted as we crunched down into the sand. Soft sand. "Uh-oh," we all reacted. The vehicle fishtailed, slowed a bit, then lurched on, and we careened up and over the next dune. And the one after that. The next was much higher, and just short of its crest we hesitated for no more than a split second, not sure of what lay beyond. With a heart-sinking whir, the Discovery's tires spun out of control.

"Whoa!" we shouted in chorus. We got out. We were dug in up to our hubcabs.

"Someone call nine one one," Ron suggested.

In what was to become an oft-repeated routine, we lowered our tire pressure to sixteen pounds per square inch. We then shoveled as much sand as possible out of the way before jacking up the rear wheels, which we then dropped back down onto aluminum sand ladders. The five-foot ladders gave us just enough of a run to send us on our way.

By radio we advised the vehicles following us when to follow our tracks and when to take a longer, easier route. We never knew quite what we were in for, especially when we crested a ridge. On the other side, we could very well slide down into a "dune pocket," a bowl of sand so steep-sided that a vehicle, even with the aid of a winch, could never climb out and would have to be abandoned.

Our route was not all up and down. On the outskirts of the Rubʻ al-Khali, dune fields alternate with gently undulating *ramlats*, or sand plains. We cruised along the southern edge of the Ramlat Mitan, and by midday we reached the dry lake once fed by the Wadi

Mitan. The temperature was in the comfortable low eighties, the air still, the day clear. We scanned the way ahead with binoculars. Somewhere out on this lakebed was a stretch of the Ubar road, discovered in 1930 by Bertram Thomas but overlaid now by recent tracks. If necessary, we could return later and seek it out. But for now, our plan was to keep moving. By day's end we hoped to intercept the road where it was well defined. We were equipped to survey and follow it for two, possibly three, more days.

Our Discoverys headed out across the lakebed, aiming for a cluster of dunes identifiable on our space imagery. It was a little unclear where to go next. We picked what appeared to be a fairly obvious route west, which led us into a winding dune street a hundred or so yards wide between parallel lines of dunes. We made good time. The dunes became higher and higher — so high they could no longer be crossed. We felt increasingly uneasy as the dune street angled us more and more to the north, farther and farther away from where we wanted to go.

On we went, flanked now by walls of great red dunes more than six hundred feet high. Nearly three hours after leaving Wadi Mitan, shielding his eyes against the low but still intense sun, Ron said, "Uh-oh," followed by, "Rats!" Ahead, the way was blocked by a massive wall of sand. We had driven into a huge cul-de-sac.

And it was us, not it, that was in the wrong place. We were lost.

Distracted by the challenge (and fun) of driving the sands, we hadn't stopped to fuss with compass bearings and land navigation. We had not paid heed to the fact that in this desert, as in all deserts, *everything looks alike*, and there is little or nothing (such as buildings, telephone poles, trees) to lend a sense of scale. From a distance, a small dune looks just like a huge dune. And features like slopes, ridges, and gullies are replicated over and over. With your nose to a map — or a space image — you can go for miles quite certain that you are where you're not, only to realize the error of your ways when you come up against an inescapably distinctive landmark. Case in point: the wall of sand before us.

We decided to camp where we were for the night. The next day, though we could ill afford it, we might have to retrace our route back to the Wadi Mitan and try another route west. We pored over a detailed Landsat 5 / SPOT image in search of our cul-de-sac. With a yellow grease pencil we marked three possibilities. None was even close to where we wanted to be.

That evening, Ron used three Kit Kat bars to illustrate that at any given time at least one NASA navigation satellite would be overhead. "But why the silence? I just find it hard to believe the whole system is down. What of ships at sea? What of animals with transmitters around their necks?"

"And what about people lost in the desert?" Kay added.

"There *has* to be some emergency provision," Ron convinced himself, and reached for the receiver, to be greeted again by " N O S A T S   F O U N D ." He grumbled and punched away at the keypad. "Aha!" he finally said, for he had discovered an advisory: in our part of the world, the system would be up and running once a day, between 2 and 3:30 A.M.

"Encouraging," Ron said, "provided these dunes don't block any signals." The Rub' al-Khali's darkly encircling dunes could easily stand between us and a satellite hovering low in the sky. Setting the receiver on the roof of a Discovery, Ron programmed it to switch on at the appropriate hour and automatically record our position.

Sunday, December 15. Day 3: searching for ghostly cities of the mind.
2:15 A.M. I woke and looked over to see that Ron Blom was also awake and up on the roof of the Discovery, "just checking" on the satellite receiver, he whispered down. "It's okay. We've got a position — 18 degrees 59 minutes 16 seconds north by 52 degrees 32 minutes 16 east."

"Good! And good night."

"Good night."

A few fitful dreams later, it was 5 A.M., time to get moving. Everyone was soon up, and with dawn still an hour away, Ron unrolled our Landsat 5 / SPOT image on the hood of a Discovery. By flashlight we saw that we were about as far away as we could be from intercepting the road to Ubar.

"We know we're up here, by this dune," Ron explained, "and where we want to be is all the way over here. And it's roughly thirty kilometers between the two, but we can't go straight there. We're going to have to work our way back down this dune street, then across to here, then strike out across this rather confused area aiming for here . . ."

With his finger he traced a route weaving through a maze of dune streets. Inevitably, though, we would have to tackle the dunes themselves. If they were anything like what was around us, they could easily be too much for us. Ran summed up our prospects: "If Ron's doing his dead reckoning navigation very carefully, shouldn't be any bother. But when you come to these two enormous lines of heavy dunes, I can't see a way through."

We began by backtracking twelve kilometers to a junction that took us into a parallel dune street. "From looking at the image, this is the only way in," Ron dryly noted. "Short of walking, that is."

We navigated very carefully now, by old-fashioned dead reckoning. Every few kilometers we would stop and set a new course. On our space image, Ron would measure where we had been and plot where we should go. I would get clear of the vehicle's magnetic field and take a compass bearing. At the wheel, Ran would hold to that bearing and track our progress in tenths of a kilometer. A single mistake and we would be lost again.

By noon, we had taken more than thirty bearings and were still apparently on course as we approached our first big line of dunes.

They were wide but not high, and we found a workable way across. We dropped into a pristine dune street, no tracks at all. We were beyond the range of wandering bedouin, drug smugglers, and military patrols.

If we could only cross the next line of dunes, we would be on the road to Ubar, close to where Bertram Thomas thought the city lay buried. At the foot of what on our space image appeared to be the most promising way across, we stopped and, with binoculars, surveyed a saddle several hundred feet above us. Mr. Gomez passed out a round of Kit Kats. We decided to give it a try with one vehicle, then have the others follow if the first made it through. Ran, Ron, and I circled to get a running start.

"Really, our only choice . . ." said Ron.

"So it's up and over or not at all," added Ran, as he drove straight into and up the dune. It was steep. It was soft. We slowed from fifty to forty to thirty miles an hour, then held at a little over twenty. I looked back. Our tracks were a foot deep. Juri, Kay, and Mr. Gomez waved us on. We climbed higher and yet higher. What did we think we were doing?

A verse of bedouin doggerel had one answer:

> Only a fool will brave the desert sun
> Searching for ghostly cities of the mind.
> Allah protect us from djinns and fiends,
> Spirits of evil who infest the dunes.[3]

"Hold on back there," Ran shouted, not quite in time to forestall my head bouncing against the roof. The way ahead now was waffled, moguled, and still steep. Ran spun the steering wheel hard one way, then the other. We slalomed onward, upward. In his shift-happy, foot-to-the-floor way, Ran drove magnificently. And we were able to radio back: "We're through! Come ahead."

It would be hard to imagine a grander or wilder or more magical desert scene than the valley, shaped like the crescent moon, that swept away below and before us. The dunes enclosing the valley were monumental, of exquisite form and color. Burnt sienna, ocher red. On its floor we spied what we thought was a fragment of the Ubar road. We should be able to see it for certain from a sand ridge across the way.

The second and third Discoverys caught up with us. How fortunate we were. Who, if anyone, had ever passed this way and gazed upon what lay before us?

With ease, we dropped down onto the valley floor. A mile farther on, Juri's voice came over the radio: "You know what you clowns just did?"

Ran answered, "No. We don't know what we clowns, as you call us, just did."

"You drove right through an encampment, that's what."

We stopped, and all walked back to where Juri pointed out a random assortment of rocks. "That? An encampment?" Ran asked, not at all convinced.

"Was once," Juri affirmed, as he began picking up and examining small stones. The first half dozen he threw over his shoulder, noting them to be worthless AFRs.[4] But then he said, "Look here now, here you've got a potsherd, though not much of a potsherd." It was orange, badly worn. It was quite old, he thought, dating to as early as 1500 B.C.

Ran examined the shard and asked what other pottery had been found in the Rub' al-Khali. Juri hesitated, then answered, "There hasn't been any, really . . ."

"What?" Ran blinked. "So this is a first piece of pottery?"

Juri believed it was. Ran shook his hand, impressed that we had an archaeologist who was "not just another pretty face." Juri chuckled and pointed ahead to more rocks, laid out in a large rectangle. He walked through a gap that could have been an entrance and prowled about, looking for more pottery or other artifacts. There were none.

He guessed that what he had found was the foundation of a brush corral, evidence that caravans had camped here.

We drove on to the far side of the valley, and on foot climbed the steep sand ridge we had spotted from the pass. We were rewarded with a panoramic view of the road to Ubar. The great track, as wide as a ten-lane freeway, emerged from under a line of dunes opposite us, crossed the valley, and was swallowed once again by the sands.

Long ago, before the road had been claimed by the sands, a great cloud of dust would have risen from the far horizon, sent skyward by hundreds upon hundreds of camels moving at once. Wary of marauders, outriders with long lances would have kept the animals in close ranks as they indignantly bleated and gurgled. They would have slowly approached where we stood and passed on by, bearing frankincense north to the great markets of the ancient world.

We camped by the Ubar road at the north end of the valley, at the edge of the L-shaped formation we had checked out on our reconnaissance, which had proved to be an ancient lakebed. Ron walked out across it and, with a hand auger, took a coring of sediments that could later be used to date when the lake had formed and flourished. His educated guess was that it had dried up sometime between 7000 and 8000 B.C.

Juri scanned the shores of the lakebed and wondered aloud, "If I came here to hunt and maybe fish, where would I camp?" "Higher ground," he answered himself, "where I could spot game and enemies." With that, he was off.

An hour later Juri was back, his every pocket clinking and bulging with rocks. No more than two hundred feet away, just out of our sight, he had found a large Neolithic (from 5000 B.C. on) campsite. He couldn't be sure, but it appeared to be divided by walkways. Scattered everywhere, broken and intact, were the utensils of Stone Age life, as many as ten thousand of them. Axe blades, animal skin scrapers, mauls, and arrowheads.

"But Stone Age," I wondered, "wouldn't that be . . ."

"Yes," he completed the sentence, "too early for what we're look-ing for."

As the moon rose and Mr. Gomez served us "Apricots, dried" and "Cookies, 2 choc. chip," we discussed the finds of the day and lis-tened as archaeologist Juri and geologist Ron pieced together a rough chronology for the valley . . .

Perhaps seven thousand years ago, Neolithic hunter-gatherers had camped on a rise overlooking what was then a small lake. Consider-ing the abundance of artifacts Juri had found, it had been a favored stopping place for hundreds, even thousands, of years. But when seasonal rains no longer reached this far inland, the lake dried up, and early man moved on, possibly to the south. The land, once savanna, became desert. The windblown sands of dry lakes and rivers formed dunes — small at first, then larger, ultimately enormous.

Sometime before 1500 B.C. (the approximate date of Juri's orange potsherd), a more technologically advanced people — almost cer-tainly the People of 'Ad — passed this way but didn't linger, other than to build simple shelters and corrals. The valley was a rest stop on the Incense Road.

"What about Ubar?" Kay asked.

There was a considerable pause, then Ron broke the silence. "To me, it comes down to water. No water, no city. There's certainly no water out here now, and frankly, I doubt that there was three or four thousand years ago. Considerably before that, yes. But when lakes like this dried up, that was it." Juri nodded in agreement and pointed out that if the region's lakes had been spring-fed rather than depend-ent on rainfall, early man would have followed the springs down, dig-ging them out as the water table dropped. That is how springs be-come wells. One or two almost certainly would still be in use.

We discussed the practicality of a city out here. Uncertain, shifting sands and violent sandstorms would have been a problem. Beyond that, what would have been a city's imperative? If Ubar was a staging point for caravans and a trading city — an "Omanum Emporium" — what was it doing sixteen days by camel from the incense groves? Out here, Ubar's control of the incense trade would have been shaky at best. Judging from our space images, there already would have been at least two opportunities to bypass such a settlement and avoid the tolls and tribute that the ancient Arabians were fond of extracting.

Simply put, a city out here would have been an economic disaster.

For the last few days, our conversation had been determinedly on the light side. We now knew why. Humor had kept us from facing the fact that we might well be chasing, as the bedouin doggerel described it, "a ghostly city of the mind."

Though we were all tired, nobody turned in for a while. The valley had spoken to us and told us what we didn't want to hear. But it nevertheless had affirmed — with a bit of orange pottery and an impressive road — that the people we sought had passed through. How and where had they begun their journey? Answer that, and we might answer the mystery of Ubar.

The valley also showed us that the Rub' al-Khali was not, as it has often been called, *natura maligna.*

The Arabs once believed that the stars were the lamps of thousands of angels. They shone brightly now, as did the crescent moon. Every curve of every dune was thrown into relief, cool blue upon dark blue. In its stillness, the valley inspired not fear, or even uneasiness, but serenity.

There is a little-known alternative translation for the phrase "Rub' al-Khali." Though it has been taken to mean the Empty Quarter since at least the 1400s, it may once, far longer ago, have meant Moon Quarter. The ancient Arabians associated different territories with different gods. The Arabian sands, then, would have been the realm of the moon god, ascendant and paramount among the gods. Rising to the sigh of cool breezes, the moon spelled relief from the heat of the day and was a lamp for caravans moving by night. The moon presided over the stars, which in turn foretold the destiny of men and nations.

By the moon's waxing and waning, all time was measured. All birth, life, and death. Long ago, an invocation cited the moon as . . .

> . . . a creature of night to signify the days.
> May the dead rise and smell the incense.[5]

Monday, December 16. Day 4: the road to — or from? — Ubar. We followed the Ubar road beyond our valley and deeper into the dunes. Our Landsat 5 / SPOT composite image was very helpful; we could cruise across the sands directly to a "blowout," a place where the road lay exposed for no more than a few hundred feet. Juri suspected that if we looked long and hard enough, we would find more evidence of incense caravans and their campsites. And, judging from the lake-beds dotting our space imagery, we would find an abundance of even earlier Neolithic sites. The idea of Neolithic sites led Juri to speculate on why the bedouin believed Ubar lay hidden out here in the dunes.

"Say you're a bedouin of the last century or so, and you find a Neolithic artifact, like a big grinding stone, which can be pretty impressive. Aha, you think, you're on the outskirts of Ubar! And your imagination gets all fired up thinking of the treasure that must be hidden under the next dune, or the next one after that. And so not only the bedouin but explorers like Bertram Thomas and Wendell Phillips get to thinking this is where to find Ubar."

Late in the morning of that fourth day in the dunes, we reached a point of no return. Getting lost two days ago had cost us considerable fuel, and now we had just enough to make it back to the beginning of the Rub' al-Khali, where we had dropped off a reserve 55-gallon drum of gas.

Reluctantly, we turned back on our tracks and, without incident, crossed the dunes that guarded our lost valley. We were then able to find a more direct route back to the Wadi Mitan. As we drove, we talked back and forth by radio. What next? Our best (and about only) hope was that we had found the Ubar road, but that it was not the road *to* the city, but *from* it. Ubar might lie in the direction we were now heading, in open desert closer to the incense groves. This was logical, but it was also unlikely, for there was hardly an inch of the open desert that hadn't been crisscrossed by sharp-eyed bedouin who, we had found, were perfectly willing to share the secrets of their land.

After the Wadi Mitan we knew we were more or less following the Ubar road, but our earlier reconnaissance had told us that we would have a hard time making it out. According to our space images, sometimes we were right on it, sometimes a few kilometers to the north or south. We considered how and where we might look for Ubar. A good start would be to explore where the road crossed wadis that once might have provided a water supply. We could also, centimeter by centimeter, again go over our space images. Had we overlooked any promising anomalies? At this point we doubted it.

It was after dark when we made it back to our first Rubʿ al-Khali campsite. Our fuel drum was exactly where we had left it, but empty — a blessing, we guessed, upon a passing bedouin's pickup. No matter, we had enough gas to make it on to the little oasis at Shisur, and maybe even back to the airbase at Thumrait.

Tuesday, December 17. Day 5: to Shisur. Desert winds can drive you to distraction. Or you may pray for them. The next day was hot and deathly still. The Discoverys kicked up huge clouds of sand that just hung there. Only the first vehicle had a view of where we were going; the others followed blindly. An hour or so out, we stopped to regroup. The cloud that had enveloped us cleared. Across the sandy plain, not all that far away, shining white buildings and towers floated in a mirage.

"Must be Ubar," Kay remarked, not very seriously. "How could we have missed it?"

Ran steadied his binoculars. "It's a housing development, would you believe," he said. "Tsk, tsk." Actually, it was the settlement of Shisur, site of the ruined fort we had seen on our reconnaissance and a dozen brand-new houses and a mosque that the government had recently completed for the principal sheiks and families of the local Rashidi. As we drove on, the mirage melted, and we could make out little kids darting between houses and alerting everyone to our arrival.

Through the desert telegraph, Shisur had heard we might be coming, and its thirty-six souls, from wide-eyed infants to white-bearded elders, all dressed in traditional Omani robes, turned out to greet us. They offered warm *salaam aliechems*; for whatever peculiar reason the strangers had come here, peace be upon them. Baheet ("Luck") ibn Abdullah ibn Salim was the imam, the religious leader of Shisur. He and his friend Mabrook ("Congratulations") proudly toured us through the newly built settlement and accompanied us as we took another look at the site's ruined fort. They confirmed that it had been built in the early 1500s by one Badr ibn Tuwariq.

The dominating feature of the ruined fort was a tower, and with more time to examine it, Juri was struck by a curious feature. Near its top, the quality of the masonry became slapdash. And the shape of the tower changed from square to round.

"You know, it could just be that this Sheik Tuwariq didn't build the fort, but *rebuilt* it," Juri remarked. "The original structure could be medieval, even earlier."

The fort was perched on the edge of the distinctive steep-walled sinkhole that gave Shisur its name. In Arabic, we were told, *shisur* meant "the cleft." Geologist Ron and archaeologist Juri led the way

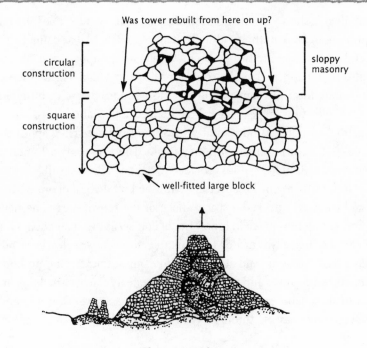

Was tower rebuilt from here on up?

circular
construction

sloppy
masonry

square
construction

well-fitted large block

The ruin at Shisur

as we walked down a sloping rubble ridge to the sinkhole's sandy floor. After some discussion, they determined that we were in what had once been an underground cavern. More than likely it had been filled with water. But at some point in the past, either through natural causes or human use or both, the water table had dropped. Emptied of water, the cavern became geologically unstable — and collapsed. Moreover, it had collapsed *after* the fort had been built. Looking up, you could see where a wall connected to the fort had sheared off, tumbled into the sinkhole, and lay buried in the sand beneath our feet.

In myth, Ubar had been destroyed in a great cataclysm whose exact nature was unclear. Different tales had spoken of a great wind, a "divine shout," or the city sinking into the sands. The Shahra tribesmen back in the mountains had told us that Ubar came to an end when "the city turned over." Could what happened at Shisur also have happened at Ubar? Might Shisur be Ubar? For that to be possible, the ruins here would have to be more than five hundred years old. Our hopes rose when Baheet and Mabrook led us to petroglyphs etched on the far wall of the sinkhole. They appeared old, but as Juri pointed out, they might date back only a hundred years or so. Out here, until very recently, time had stood still.

Returning to our Discoverys, we finished off the last of our MREs and carefully checked a space image of the Shisur area. The new houses and mosque didn't show up, of course, as they had been built after the image was made. The sinkhole, though, was clearly visible as a dark crescent. And at least six old caravan tracks came up from the incense groves and converged on the site. None bypassed Shisur; with its reliable water source, it was a necessary stop for any and all caravans passing through. Whoever controlled Shisur — and its

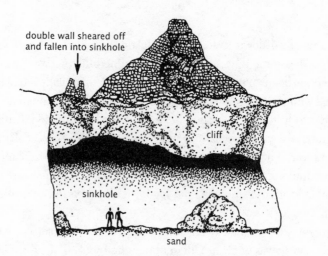

Shisur's sinkhole

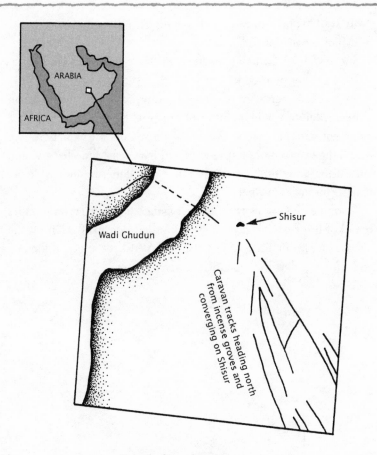

Wadi Ghudun

Shisur

Caravan tracks heading north from incense groves and converging on Shisur

Detail of Landsat 5 image

water — could control the incense trade as the caravans headed out across the Rub' al-Khali.

There were, of course, arguments why Shisur couldn't possibly be Ubar. So far as we knew, the fort was not old enough by a good two thousand years. And the site was hardly a candidate for the city described in the Koran, a city "whose like has never been built in the whole land."

"It wouldn't take all that much, would it?" Kay asked.

"Much what?" I asked.

She explained, "Much to be the greatest city in all *this* land."

"Perhaps not," mused Juri.

We cracked open a box of Kit Kat bars and shared them with Imam Baheet, Mabrook, and a cluster of Shisur kids. And we agreed to a plan. For a month we would make Shisur our headquarters. We would dig two or more test squares and try to date the site's rise and fall. At the same time we would range out along the incense road, looking for traces of Ubar.

A couple of Kit Kats later, Baheet agreed to a price for a month's rental of three of Shisur's not-quite-finished houses. We left for the airbase at Thumrait, looking forward to real beds and real food — and, after five days in the sands, showers.

15

What the Radar Revealed

WE SOON RETURNED TO SHISUR. We had reinforcements. Joining us now were JPL's Charles Elachi and Kris Blom, Ron's wife, also a JPL scientist. First on our agenda was to probe the sands that had drifted into the site's yawning sinkhole over the centuries. We would do this with a mini version of the radar that, seven years before, had scanned this desert from the space shuttle *Challenger*. Our three JPL scientists unpacked five crates containing the components of a Geosystems SIR ground-penetrating radar rig. With stakes and string, Juri, Kay, and I laid out a grid on the floor of the sinkhole. Ron assembled the radar's sender-receiver — a red sled resembling an oversize carpet sweeper — and connected it by cable to a stationary signal processor and recorder. Kris switched it on. Graph paper rolled; ink flowed. Charles set the unit to record what lay under our feet to a depth of fifty feet.

We were all set to go when, without warning, the whole array went dead. Charles checked the power supply; Kris leafed through the manual; Ron thought the recording unit might be overheated. The three scientists moved it into the shade and waited for it to cool off. Still it wouldn't run. They tried everything. "Myself," Ron muttered with mock disenchantment, "I still think the best way to find buried objects is to dig them up."

Everyone was about at wits' end when Ran Fiennes ambled over and politely asked, "You mind if I try something?"

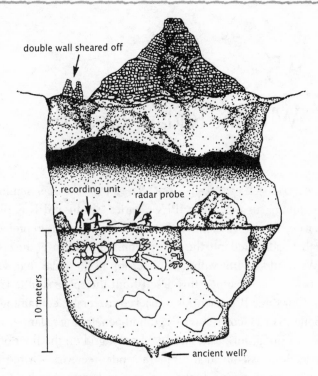

double wall sheared off

recording unit   radar probe

10 meters

ancient well?

Shisur's sinkhole as radar mapped

"Not at all," Ron said.

Ran bent over and lifted the recording unit a foot in the air. And dropped it. Sixty thousand dollars' worth of sophisticated hardware hit the ground and whirred to life. Looking on, Baheet ("Luck") and Mabrook ("Congratulations") uttered, "Hamdullalah" — thanks be to Allah.

Ron was soon dragging the red sled methodically back and forth across the sinkhole. Bedrock was thirty feet down, and between us and it, the radar detected a jumble of fractured rock and, quite possibly, fallen walls and broken buildings.

Charles and Kris monitored the recorder's complex squiggles. A sudden change caught Charles's eye. "Look here," he said, "see how it comes down?"

"Which means?" Juri asked.

Charles thought a minute and replied, "A well. I would say there's an old well down there." Further scans confirmed it. In the center of the sinkhole was the shaft of a well that might have been sunk by the People of 'Ad as, faced with a diminishing water supply, they sought to save their desert stronghold.

That night Juri was drawn back to Shisur's fort and sinkhole, and by flashlight he walked the site. Kay and I tagged along. Someone had gone to a lot of trouble out here, quarrying and dressing thousands of stone blocks to make their stand in this remote desert wilderness. It would be nice to say that Shisur's stones spoke to us. But they didn't. As Juri pointed out, nothing belied their age. They could date to the days of the incense trade or, as most everyone thought, they could be the handiwork of the Yemeni sheik Badr ibn Tuwariq in the 1500s.

"Time to stop speculating," Juri said, "and start digging."

16

City of Towers

FROM TIME TO TIME, Imam Baheet would ascend the minaret of his settlement's new mosque and summon the faithful to prayer . . .

> God is the greatest,
> There is no god but God.

Baheet's call rang out across the tiny settlement and its nearby ruins. How strange it would be if Ubar lay buried within sight and earshot of where the faithful gathered to chant *suras* (chapters) from the Koran, suras that proclaimed:

> Arrogant and unjust were the men of 'Ad. "Who is mightier than we?" they used to say. (from the sura "Revelations Well Expounded")

> Have you not heard how Allah dealt with 'Ad? The people of the many-columned city of Iram, whose like has never been built in the whole land? (from "The Dawn")

> On a day of unremitting woe we let loose on them a howling wind which snatched them off as though they were trunks of uprooted trees. (from "The Moon")

> And when morning came there was nothing to be seen besides

their ruined dwellings. Thus we reward the wrongdoers. (from "al-Ahqaf")

'Ad denied their Lord. Gone are 'Ad . . . (from "Hud")[1]

We settled in at Shisur. The three houses we had rented were filled with fine red sand that had to be shoveled out, our first excavation. Returning from a run to Salalah and the coast, Kay arrived with every inch of her Discovery packed with pink and turquoise foam rubber mattresses that she had found in the souk for four dollars each. When she opened the door, they sprang out and flew all over the place, much to the amusement of the locals. For a homey touch, Kay distributed brightly colored cotton bedspreads that featured sayings in an unknown (to us) language: "NAMI KAMA MAMA." "ADUI NI MDOMO WAKO." Swahili, Juri thought.

From the airbase at Thumrait, Ran procured a woebegone generator and coaxed it to life. It gave us a couple of hours of evening light and powered our Racal radios, our link to the airbase and the world beyond. Curiously, the world beyond seemed remote from us, not we from it. We were quite happy just to check in with Thumrait twice a day, at one and seven P.M. In a typical call we discussed the impending arrival of five student diggers from the States and were delighted to learn that Airwork, as a gift, would be providing us with two twenty-pound Christmas turkeys.

Juri's Southwest Missouri State students were a little dazed when they arrived at Shisur, about as far away as they could be from the rolling cornfields of home. Three were undergraduates: Rick Brietenstein, Jean England, and Julie Knight. Rick had never been outside the country or even on a plane, yet he would soon be hard at work, both digging and compiling a tribal *Who's Who* of Baheet, Mabrook, and related Shisurites. We had two grad students, assistant archaeologist Jana Owen and registrar (recorder and keeper of our finds) Amy Hirschfeld. Both were experienced in the Middle East, having worked together in Israel. Kay and I were also delighted by the appearance at Shisur of our daughters, Cristina and Jennifer, on Christmas breaks from work and college (Cristina was editing a magazine; Jennifer was in her last year at Wesleyan University).

To dig Shisur, Juri came up with a simple, specific plan: survey and test-excavate the ridge running east from the old fort (see plan, opposite). The ridge didn't look terribly promising. A few meters beyond the fort the broken walls of three small rooms rose from the dirt and sand, but otherwise the ridge appeared to be a purely natural feature edging the north side of Shisur's sinkhole. Juri explained that if this indeed was true, we could focus on the ruins of the fort and wrap up work here in relatively quick order. But if the ridge produced results, the site might amount to something.

For surveying purposes, Juri and Jana Owen established a zero point, then measured everything out from there. Sighting with a theodolite, they laid out a grid of three-meter by three-meter squares and marked each with metal rods and orange string.

While Juri and his crew prepared to dig the ridge, Kay bounced off across the desert in a Discovery and returned triumphantly with a bone-bleached, leafless bush in the passenger seat. Our Christmas tree. To the accompaniment of carols played on little speakers connected to her Walkman, she decorated it with a strand of twinkly lights, which she'd found in a hardware store on the coast.[2]

On Christmas Day, Mr. Gomez outdid himself. As Kay had promised, he now had real food to cook, and though refrigeration was impossible, he had a two-burner butane stove supplemented by desert stoves improvised by our Airwork volunteers. To make one you fill a metal ammunition case with sand, then riddle it with automatic rifle fire. Pour in gasoline, toss in a match, and you have a hot, smokeless (in case of lurking enemies) fire. Today a dozen Airworkers joined us, bringing gifts of Christmas plum pudding, brandy, and their good company.

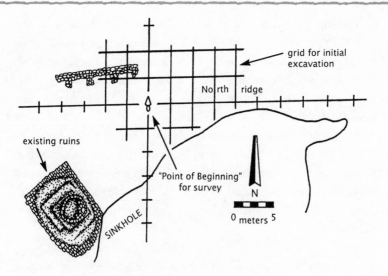

North ridge 1: before excavation

We found a place for everyone at one long table set outside. Our seats were concrete blocks borrowed from the construction of new Shisur. Our American students, all away from home at Christmas for the first time, took turns describing their family gatherings, and our Airwork volunteers, committed to an eleven-month stretch without leave, talked of snowy traditional Christmases in England and Scotland. Juri reminded us that the first Christmas had more to do with palm than pine trees and that the gifts were "gold, frankincense and myrrh."

"There's a good possibility that's translated wrong," he added. "Gold could mean not shiny metal stuff, but a 'gold grade' of incense, perhaps a balsam. The Bible tells us there were twelve, maybe more, kinds of incense. So the gifts could have been what you'd find not far from here: three kinds of incense. Golden balsam, silver frankincense, and myrrh."

If Shisur proved to date to biblical times, incense caravans may well have set out from here on a long and arduous journey north across Arabia — and, for some, on to Jerusalem. In order to return home before the scorching heat of summer, Arabian traders would have timed their arrival in Jerusalem for late December or early January.

Across the valley from Jerusalem are caves where the traders might have sheltered their camels from the winter rain and cold. Often, when local inns were full, traders and other travelers stayed in the caves. If an infant was born in their midst, Arabian wayfarers would have considered themselves blessed and offered the child gifts of their incense.

Week two at Shisur . . . We dug. Slowly, with trowels and brushes. Excavation wasn't a process to be rushed. If there was anything here, it would be revealed in good time — and it was. At depths ranging from a few inches to a few feet, Juri and his students uncovered the stone foundation of what was once a wall. It ran along the ridge that had appeared to be a natural feature. The three small rooms Juri had noted backed onto the wall. He speculated that they could be store-rooms or merchants' stalls: "In souks all over Arabia, you still see shops laid out like this."

Our student diggers were each assigned a three-meter square. As they carefully excavated, they recorded the positions of rocks — some of which were clearly the foundations of a wall — and noted subtle changes in the composition of the sand and dirt. Their initial modest finds included bits of worn orange pottery and tiny bones (probably mice). Juri circulated about, answering questions ("Is this worth anything?" "No"), making suggestions ("You can pull those little rocks out. Not structural"), and often pitching in with the spade-work.

Four days after Christmas, poking about in an untouched square,

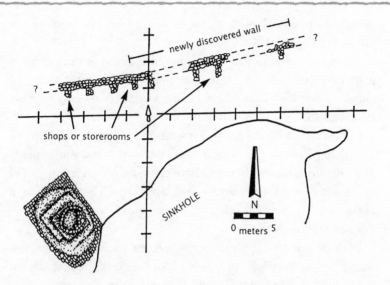

North ridge 2: wall revealed

Juri unearthed a shard. Easily overlooked, it was dull gray, a contrast to the orange ware he and his students had been finding. Picking it up, he turned it over, then over again. "Nice early piece" was all he could say, for he was stunned. This "nice early piece" was a fragment of a Roman jar, either brought here by caravan or copied in Arabia as "imitation ware." In either case, "early" meant before the time of Christ, possibly as early as 300 B.C.

Excavation intensified. To be sure that nothing was missed, each square's sand and dirt was collected in black plastic buckets and carted over to a sifting screen, where it proved, more often than not, to be sand and dirt, nothing more. Much of the screening fell to Absalom, one of three enterprising Baluchi laborers who had queried Ran at the coast, made their way to Shisur, and been hired. When a student spread the contents of a bucket on Absalom's screen, he shook it only briefly before answering repeated inquiries of "Anything? Anything yet?" with a dark Baluchi frown. Archaeology was not for the impatient.

It was student Julie Knight who found the next distinctive bit of pottery. Without reference texts, Juri couldn't be sure what it was, but guessed it was Greek (or imitation Greek), datable to 100 B.C. at the latest and 400 B.C. at the earliest.

In the coming days, a few shards were to become hundreds. More Greek and Roman pieces, and some that Juri could not immediately identify but thought might have come from the eastern part of the classical world, from Syria or perhaps Persia. The settlement at Shisur was no longer five hundred years old; it was well over two thousand years old!

With its varied ware, Shisur began to reveal its past. Its inhabitants must have prospered, for they could afford some of the best utensils the ancient world offered. Beyond that, they were themselves inventive, producing orange pottery decorated, frequently, with a motif of dots inside circles. Juri had found such ware at Khor Suli on the coast, probably out in the Rub' al-Khali (a piece so worn he couldn't be sure), and now here. He believed the style was unique to this part of Arabia — and possibly a hallmark of the People of 'Ad.

Now we dared whisper, "Ubar?" Kay, raised in the South and familiar with things like jinxes and spells, said we should be careful

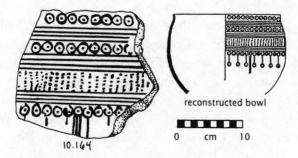

10.164

reconstructed bowl

0   cm   10

Dot-and-circle shard

and not risk spooking our good fortune. If at this point we went around thinking we had found Ubar, it could somehow make the place *not* be Ubar. If what we had found was too good to be true, maybe it wasn't. And there was now a nagging question: was our site a backwater outpost — or was it, as Ubar must have been, a significant and major settlement?

Week three at Shisur . . . Monday passed without incident. On Tuesday the wall that appeared to extend east from the existing fort puzzled Juri. In Rick Brietenstein's square, instead of continuing straight on, the wall made an unexpected curve. "Comes off clean," Juri puzzled, "and curves around." He and Rick followed the wall stone by stone, questioning whether they were being deceived by collapsed masonry or, worse yet, a natural line of rocks. But no, the wall was distinctly there, curving around like a horseshoe, then abruptly resuming its prior alignment.

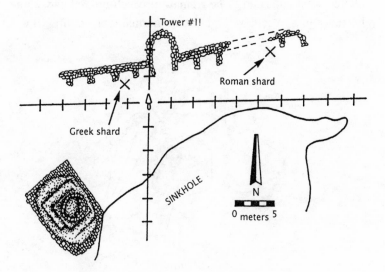

North ridge 3: tower discovered

Juri straightened up, stepped back. And it came to him. "You know what? A tower. Looks like we have ourselves a tower."

We clambered up on the roof of a Discovery for a high-angle view of the excavation. As Amy Hirschfeld focused her Nikon, Juri had Rick chalk an ID slate: "DAD [Dhofar Antiquities Dept.] T O W E R #1." From the width of its stone foundations Juri estimated that the tower might have risen as high as thirty feet.

"A tower, think about that," Juri said. "You don't just build one in the middle of nowhere. You have a wall here, then a tower, then you're going to have more wall, more towers . . ." Here at Shisur a tower would almost certainly have been a component of a large structure: a fortress that protected the site's water supply and guarded a season's store of frankincense.

Sure enough, almost simultaneously, farther down the ridge Jean England unearthed in her square the foundations of a second tower, "DAD T O W E R #2." Larger than the first and circular, it marked the fortress's northeast corner. It contained traces of an interior stair and sheltered what appeared to be a small furnace outfitted with stone reflecting vanes to achieve higher than normal temperatures. It was

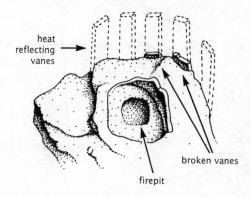

Ancient furnace

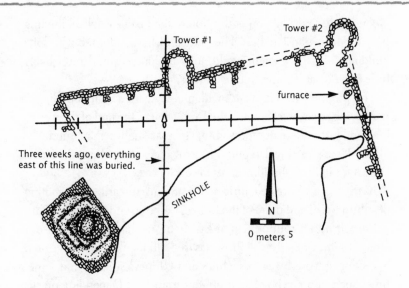

North ridge 4: excavation completed

difficult to say what the furnace had been intended for. It wasn't for smelting metal, for there was no evidence of slag. It might have been used, we thought, to process frankincense.[3]

Juri's hunch to excavate the site's north ridge could not have been more on target. We had uncovered the north wall and towers of an ancient fortress.

As, at dusk, Baheet issued a call to prayer from Shisur's minaret, I was prompted to read, as I had read many times before, the Koran's sura "The Dawn" . . . "Have you not heard how Allah dealt with 'Ad? The people of the many-columned city of Iram, whose like has never been built in the whole land?"

If this was Iram/Ubar, where were the columns? One explanation, I thought, lay in the Arabic word عماد , pronounced *imad*. In contemporary usage, it means pillar, but older definitions were broader. In

George Sale's 1782 edition of the Koran, the first in English, the line in question is translated as "The people of Iram, adorned with lofty buildings." In the ancient world, lofty buildings would most likely have been what Juri was finding: towers.[4]

The prophet Muhammad, incidentally, decried "lofty buildings." In a saying regarding "Signs of the End" (that is, the end of the world) he condemns them for presuming to soar higher than mosques. Given Ubar's mythical repute for arrogance, how fitting that it be known for its "lofty buildings," its towers.

As the week progressed, Juri and his students unearthed the footing of a third tower and more of the fortress wall (see endpaper site plan).

We fell into a routine. Up at first light, we had breakfast in our largest room, where Kay's Christmas tree still twinkled in the corner. Mr. Gomez wore his cook's whites and Chinese slippers and sometimes sported a cowboy hat, a present from Kay. Depending on his mood, there would be cereal, pancakes, even cheese omelets.

Then what became known as "the March of Archaeology" would proceed down the village's main street. Juri's not-totally-awake students led the way, laden with buckets, notebooks, and surveying gear. Next came our volunteers, some of whom had driven eight hundred miles across the desert from Muscat to help out. The rear was brought up by our three Baluchis and their wheelbarrows. At the end of Shisur's main street — all of three houses — the March would turn left and soon arrive at the ruins, where the group dispersed to dig, screen, and take notes.

Around ten o'clock we would break for tea and Kit Kats, often joined by Baheet and Mabrook. When digging resumed, they would drift from square to square, help out as the spirit moved them, and contemplate the idea that Ubar, a site celebrated in countless generations of bedouin lore, might actually lie beneath their feet. On one occasion Imam Baheet got surprisingly worked up and proclaimed: "The People of 'Ad were corrupt. We all know it. Allah punished them!" For emphasis he picked up a large rock and thumped it down on the ground, adding "Ubar! Khalas!" (Finished!) Being an imam and warming to the lesson of the Allah-smitten Ubarites, he would sometimes make a stab at converting us. But he always allowed that Islam had great respect for "people of the Book," meaning the Bible.

By noon it was usually uncomfortably hot, and the site offered no shade. We would work as long as we reasonably could, then at one P.M. or a little after, the March of Archaeology would retrace its steps. After a light lunch, everyone would lie low for a few hours, updating their field notes, writing home, or reading Bertram Thomas's *Arabia Felix* or, for a change of scene, a dog-eared copy of Elmore Leonard's *Glitz*.

About three-thirty, with the hottest part of the day past, digging would resume. Often Juri would leave assistant archaeologist Jana Owen in charge of the site, while he ranged out across the desert, accompanied by Baheet or Mabrook, who knew its every rise and hollow. He had a hunch that the fortress at Shisur was the center of a large seminomadic settlement. Our space imaging revealed that northeast of Shisur there had once been a large slow-moving river. Neolithic man had been drawn to its banks, and when the river ceased to flow, our People of 'Ad might have camped there, for the land could still have been fertile, a sprawling oasis.[5] Drawn up around dozens of rock-ringed fire pits — still clearly visible — caravans would have prepared for the crossing of the Rub' al-Khali.

Around six in the evening, we would often join the daily majlis, or social hour, held by Shisur's Rashidi elders. Though their new houses included special majlis rooms, they were more comfortable taking their coffee around a fire laid in a cut-down oil drum set out in the main street. Their conversation often dwelt on the virtues and vices of the several camels that wandered the village. Displaced by Toyotas, the camels at present had no clear role. If anything, they appeared to be backups if the Rashidi's current rather easy life fell apart.

This was confirmed when the discussion one day turned to water. The water table in the immediate area was measurably dropping, and there was a chance that Shisur's water supply, after thousands of years, could run dry. What would the Rashidi do? The question proved a test of their bedouin spirit. With not a glance at their fine new houses, Baheet and Mabrook shrugged and looked out across the desert. "We go," Baheet said.

At dinner we would compare notes, lay plans for the next day, and enjoy our various cuisines (U.S. college / Arabic / U.S. veggie / Baluchi vegetarian), and nod our heads in sympathy when Mr. Gomez let loose with a tirade, or at least what we thought was a tirade — only Kay could tell for sure.

One night the situation got pretty serious as, eyeing our Omani police guards, Mr. Gomez said that *someone* had been in his off-limits storeroom, and *that someone* had helped himself to the brandy he reserved for special dishes.

Unfailingly courteous, our policemen had often expressed their appreciation with a "Thank you too much!" (They didn't see any particular distinction between "too" and "very.") On this occasion, Jumma, their leader, responded to Mr. Gomez's allegations with a distinctly sarcastic "Thank you too much, Mr. Gomez."

"Thank you not very much, Mr. Policeman Jumma," Mr. Gomez replied, and stalked out into the night.

Juri confirmed that it is not the wonder of the past or the fate of ancient nations that eventually becomes the major concern of archaeological expeditions; it is the food. Its shortcomings, its satisfactions. Our American students were never happier than the day when, with Kay's guidance, Mr. Gomez served up tomato soup followed by a choice of grilled-cheese sandwiches or hamburgers (odd, egg-shaped hamburgers, but hamburgers nonetheless).

It fell to Kay to calm Mr. Gomez and, without the benefit of refrigeration, make sure we had enough food for up to forty people at a time (on the weekends we hosted a legion of volunteers). Every couple of nights, after dinner, she and Mr. Gomez would take stock of what we had and what we needed. Then Kay would slide into a Discovery and drive off across the desert alone, so that she could load it up with as much fresh food as possible. She loved the desert, especially at night. She had high-powered halogen driving lights, and high- and low-frequency radios to call us in case of a breakdown. She could also tune in to the international news, on either the BBC World Service, transmitted from the Persian Gulf, or on the Voice of America. In this part of the world the VOA news in English was de-li-ver-ed ve-ry, ve-ry s-l-o-w-ly. As she heard what was going on with Microsoft and the Moral Majority, she kept an eye on the odometer. She knew, better than anyone, at what mileages to expect camel wallows, and she slued through them with ease.

The round trip to our depot at the airbase at Thumrait took three to four hours. Around midnight I would climb to the flat rooftop of our Shisur house and, sooner or later, I would see, way out in the night and across the desert, two tiny bright dots. Thirty miles away, they moved slowly, disappearing from sight, then bumping over a rise, closer, a few minutes later. Within the hour, Kay would be home.

By flashlight we offloaded the contents of Kay's Discovery into Mr. Gomez's storeroom, and this was the only occasion when the two were at odds. It was predictable; it had to do with the cases of Wadi Tanuf spring water. Tucked in each was a souvenir glass, worth about three cents, decorated in blue with the legend "Wadi Tanuf" in English and Arabic. Both Kay and Mr. Gomez were intent on acquiring a set and, with claims and counterclaims, fought over the glasses as if there were no tomorrow.

In our third week at Shisur, we radioed the news of our discoveries to Thumrait, to be relayed on to our sponsors in Muscat. The next weekend the site was thronged with volunteers. Juri deployed them along the projected course of the site's wall, assigning a student supervisor to each half dozen.

Joan Fulford, a volunteer, had been digging for no more than twenty minutes before a blue-green glint caught her eye. To her great delight, she brushed clear an exquisite Roman vase. "Me?" she exclaimed. "Oh dear, I found this? My, but those people had nice things!"

A few meters away, the idea that the site was once a bustling marketplace — deserving of the designation "Omanum Emporium" — was given further credence as Airwork volunteers Richie Arnold, Nick Deufel, Neal Barnes (Guru), and Pete Eades (Black Adder) enthusiastically attacked a slope overlooking the sinkhole. Waste rock flew through the air, and barrow after barrow of sand was hauled away to the sifting screens. Within a few hours they were rewarded with the outlines of shops, more and more of them, backing onto the site's encircling wall.

When Ran Fiennes, normally busy with logistics and government liaison, had a go at actually digging, he excavated with his hands rather than a trowel. "Now, we don't dig like that," Juri advised. But there was no stopping Ran as he hit a section of the wall and proclaimed, "I'm an archaeologist, an overnight archaeologist!"

"Stop digging like a fox," Juri pleaded, to absolutely no avail.

It was an enormously rewarding week. Juri identified a total of five towers and suspected that two or three more were hidden in the site's rubble and sand. Our fortress's encircling wall finally led us back to where we had started: the old fort, which we now called the Citadel. It was a sizable, complex structure, and dangerous too, for it was severely undercut by Shisur's sinkhole. A week before, without warning, several tons of rock had sheared off the sinkhole's south edge. If anybody had been underneath, they would have been killed. It was now quite possible that, triggered by the vibrations of people at work (or malicious djinns), the entire Citadel could come crashing down. To excavate this structure, we recruited volunteers who had had

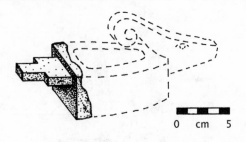

Reconstructed lamp

mountaineering or spelunking experience. They donned safety harnesses and ran ropes to bumpers of Discoverys parked out of harm's way. If the citadel collapsed, they might suffer a nasty jolt but would be left dangling in the air rather than buried under the rubble.

A few days' excavations revealed halves of rooms, meaning that the Citadel had not been built at the edge of the sinkhole, but had been larger and had collapsed into it. Excavating one of the rooms, Pete Eades discovered the Citadel's first artifact (shown above). Gingerly, Juri made his way over and turned it this way and that. At first he thought it was part of an incense burner, then realized he was holding the handle of a lamp.

A lamp that had shone as this remote settlement thrived. A lamp that fell to the floor and flickered out when it was suddenly, violently destroyed.

Week four at Shisur . . . The weather took a turn for the worse. A raw, cold wind raked the site, blowing sand in our eyes and down the backs of our necks. It was a week for spending time in Juri's workroom and Amy Hirschfeld's lab. Our initial excavations had circled the site, and it was time to clean, sort, and inventory what had been found.

Amy's lab was now piled with thousands of artifacts: bag after bag of broken pottery, beads, bracelets, glassware, and fragments of three

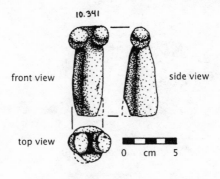

Sandstone artifact

incense burners. Each artifact was given an ID number and logged on an IBM 386 computer. Dbase 4 software three-dimensionally pinpointed the exact level and location where every bit and piece had been found. Some of the artifacts were puzzling (and some still are). In what once was an ancient shop or storeroom, volunteer Ian Brown unearthed a sandstone object, which was unusual because sandstone is not found anywhere near Shisur. Juri first guessed that the rock was a cultic object, perhaps a small betyl.

Then, across the ruin, in rubble in the base of Tower #6, five more similar-sized pieces of sandstone were discovered. Juri lined the pieces up. As he often did, he rubbed his nose and fiddled with whatever was handy, a pencil or a dental pick or in this case an artifact. He ventured an awful pun or two (singing "These stones in the foundation, what do they mean to me?").

"They seem to go together," he pondered, "as a set. You know what, Amy? I don't think these are religious at all . . . How about a queen [shown above], with her queenlike attributes? Then some pawns, maybe. I'm not sure what these others are."

Juri and Amy were looking at the oldest known chess set found in Arabia, one of the oldest in the world. As well as a queen, there were

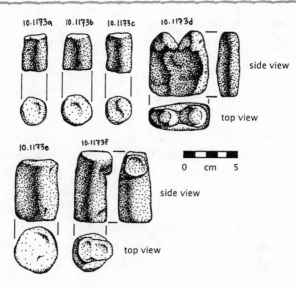

More sandstone artifacts

(left to right above) three pawns, a bishop or vizier with a telltale "hat," a castle, and a knight. Finally, discovered outside the tower, there was a king, inscribed with a six-pointed star.[6]

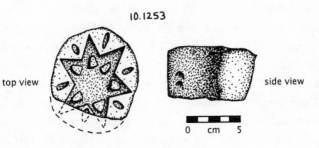

Sandstone king

The chess set, wonderful though it was, raised a further question. The game is believed to have originated in India in the late 600s, a hundred years or so *after* Ubar's legendary destruction. Did this checkmate our Ubar theories? Juri thought not. The sinkhole, he reasoned, could have collapsed sometime after 150 A.D., and the fortress thereupon abandoned. Then, centuries later, ruined Ubar may have been reoccupied.

The Ubarites did not vanish from the face of the earth after fleeing their city; rather, there is evidence that they were absorbed by other tribes (the Shahra of the Dhofar Mountains and the Mahra of the Oman-Yemen borderlands). In time, members of these tribes could have been drawn back to the ruined city and its unique water source. While their animals grazed on what was left of the verdant oasis, they could have whiled away the hours with the newly invented game of chess.

Baheet and Mabrook often dropped by Amy's lab to help out. Particularly when the sky was dark and the wind howled through their village, they would sit for hours, quietly cleaning shards and piecing them together. With the help of a magnifying glass, they translated fragments of Arabic barely visible on Islamic coins. And, forthrightly, they would ask questions that we tended to avoid. Such as: "If this is Ubar, then where's the gold?" Many bedouin assume that no matter what archaeologists say, they are treasure seekers. Why else would they dig in the dirt? Baheet and Mabrook were too sensible for that; nonetheless, they reminded us that the Ubar storied by their grandfathers was a city of ذَهَب أَحْمَر, *dhahab ahmar* — red gold.

We hadn't found even a hint of gold. One good reason for this was that we hadn't found any skeletons either. When archaeologists *do* find treasure, it is usually in the form of grave goods, offerings to comfort and aid the deceased in their afterlife. But Juri had found no burials at or near Shisur. One possible explanation was that anyone who died there was taken to the mountains for burial, perhaps in the

Vale of Remembrance that we had visited. This would have been likely, we thought, if the fortress of Shisur had been *seasonally occupied*. The People of 'Ad might have spent the late spring and summer months in the cooler, monsoon-shrouded Dhofar Mountains, harvesting their precious frankincense. Come fall they could have transported it to Shisur, a staging area for the great caravans striking out across Arabia. When these caravans returned to Shisur in the spring, the People of 'Ad would take their dead — as well as their profits (including any gold) — back to their mountain retreat.[7]

So it was that Baheet's blunt "Where's the gold?" led to the working out of a scenario of the annual ebb and flow of life at Shisur. Depending on the time of year when the sinkhole collapsed, there may have been treasure stored in the Citadel. But, Baheet and Mabrook agreed, the 'Ad, when they fled, would not have been so panicked, morose, or witless as to leave any treasure behind.

As the nasty weather allowed, Juri's students continued to clear away foundations of towers, walls, and the Citadel. It was hard to tell whether the Citadel was a stronghold, as it first appeared to be, or a temple dedicated to ancient gods — or both. We could only speculate that whoever had raised this fortress had been inspired by more than an elemental need to stave off their enemies.

In terms of productive archaeology, the next weekend was not the time for solving the inner meaning of the Citadel. The weather lapsed back into wretchedness, and only a handful of volunteers showed up. It wasn't the stinging sands, we learned, that kept them away, but the fact that an Omani fighter plane based at Thumrait had plunged into the Arabian Sea. Over the radio we offered our condolences, but were told, oh no, the pilot had ejected and was quite all right.

Our volunteers, a slightly slurred voice informed us, were partying. That was strange, we thought, for weren't our Airwork volunteers responsible for the maintenance of Thumrait's aircraft? Yes, yes, they were, and that was the point. The crash had been due to *pilot error*; a nonstop, sixty-two-hour party was celebrating the fact that *the crash wasn't their fault!*

Week five at Shisur... This was to be the last week for all but Juri and his five students.

For years George Hedges, now back in Los Angeles, had worked tirelessly to organize the expedition and get it financially off the ground. In the field, Ran Fiennes had done a superb job of orchestrating our logistics. We always had what we needed, and Juri never lost a minute as he excavated Shisur. Our original plan had called for the sinking of one or two modest test shafts if we found a site that appeared to be Ubar. As it turned out, with as many as forty people digging at a time, we had brought to light an entire buried fortress and surveyed its surrounding terrain.

Our plan had called for three months in Oman, and now those three months were about up, and our bank account was about done for. There was just enough money left to see Juri and his students through the rest of their work-study semester.

In our last week at Shisur, a radio call alerted us that an Omani Air Force Huey had refueled at Thumrait and was on its way north to Shisur. What good fortune! We could photograph the site from the air. The helicopter and its crew showed up in time for dinner; the English pilot and Omani copilot found it hard to believe that anybody had built *anything* so far out in the desert so long ago.

The next day was clear, with not a breath of wind. We were quickly aloft and circling the site. With the tiny figures of Juri's students for scale, it looked bigger than it felt on the ground. We could easily make out the foundations of several towers and see that the site had had not only an outer wall but, most likely, an inner wall to protect the immediate area of the Citadel; it had been all but wiped out in the sinkhole's collapse.

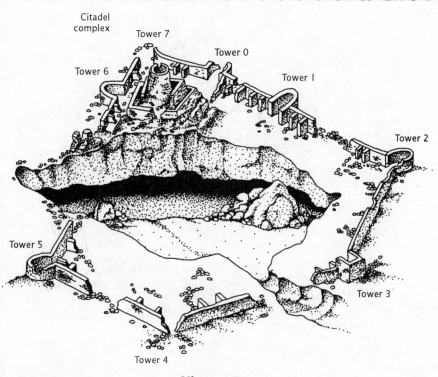

Ubar in ruins

This, we finally believed, was ancient Ubar.

For over a month now, we had worked a puzzle. Its key pieces had to do with pottery types and sequences, architectural plan, and the role of the incense trade. Along the way, some pieces didn't seem to fit. What was a chess set doing here? Where, as Baheet asked, was the gold? In finding what fitted with what, there had been no magic "Eureka!" moment, no time to proclaim "This is Ubar!" and unleash the formidable party potential of our Airwork volunteers. Instead, a picture had slowly formed of a distant time, place, and people, a picture that seen in its entirety was a convincing match for the legendary lost Ubar.

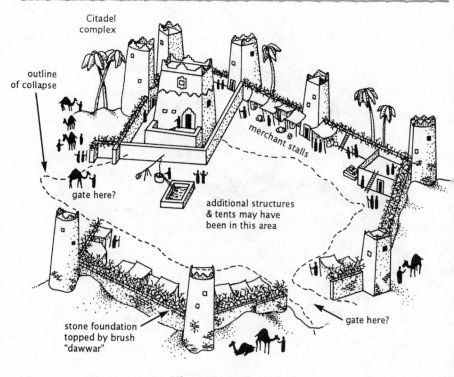

Citadel
complex

outline
of collapse

merchant stalls

gate here?

additional structures
& tents may have
been in this area

stone foundation
topped by brush
"dawwar"

gate here?

Ubar as it may have been

Key to that match were . . .

LOCATION   The site was where it was supposed to be. The
myth of Ubar had led us to the unremitting desolation of a
remote area of Arabia — and, against all expectation, an impressive fortress.

AGE   The site was ancient. In myth, Ubar was founded by
Noah's grandson, a first patriarch of the People of 'Ad. What we
had found dated to 900 B.C. or earlier — the very dawn of civilization in this land. Our site was among the oldest, if not *the*
oldest, of Arabia's incense-trading caravansaries.

CHARACTER   Here was an expression of the Koran's ذَات العِمَاد, *dhat al-imad,* city of lofty buildings. And Ubar's eight or more towers guarded a water source that, more than anything else in the surrounding fifty thousand square miles, qualified as "the great well of Wabar" described by the historian Yaqut ibn Abdallah as the city's principal feature. For all its isolation, here was a place where, as in its legend, people prospered and lived well, cooking and dining on the ware of classical civilizations.

DESTRUCTION   The legend of Ubar climaxed as the city "sank into the sands." It surely did. Ubar wasn't burned and sacked, decimated by plague, or rocked by a deadly quake. It collapsed into an underground cavern. Of all the sites in all the ancient world, *Ubar came to a unique and peculiar end, an end identical in legend and reality.*

As our helicopter, in widening circles, flew out over the desert, Juri pointed out geological traces indicating where springs once broke the surface and supported a substantial oasis. Water, he felt, was the key to our site's identity. Once there had been many springs; finally there was but one, and it still flowed. "In this desert," he later observed, "Ubar could have been hidden anywhere in, say, fifty thousand square miles. But it's here because there's water. Permanent water."

We had just enough fuel to helicopter to the northeast and photograph the nearby low hills where caravans rested before venturing off across the Rub' al-Khali. Returning to earth, we knew that, as almost always in archaeology, we could not identify the site as Ubar without reservation. We could never be 100 percent certain unless we found an inscription that included the word)Ⅲh, Ubar. Father Jamme had written this out for us, but doubted that we would ever find it.[8] We consoled ourselves with the fact that no telltale inscription has ever been found at what archaeologists have agreed is Homer's Troy. We were fortunate enough to find what we had found.

It was time now to head home. Ran and the film crew, Kevin O'Brien and George Ollen, were to fly out of the coastal town of Salalah. Kay and I would drive overland to rendezvous with them in Muscat. We packed our Discovery with equipment on loan from our Omani sponsors or to be shipped back to the United States. We said goodbye to Juri and his students; we thanked Baheet, Mabrook, and the people of Shisur. We would leave at first light the next morning.

After breakfasting with Mr. Gomez, Kay and I stepped out into Shisur's dusty main street . . . and were startled to find it lined by all our Rashidi friends. They had turned out to see us off and wish us well. We shook their hands, gave them hugs, then waved and wiped tears from our eyes as we drove away. Following the route of the March of Archaeology, we skirted ruined Ubar, then picked up a desert track heading east. "We'll be back, I'm sure we will," Kay said. "But it can never be the same, you know?" She quietly cried all the way to the oil camp at Dawqah, where we turned onto the paved road to Muscat.

17

Red Springs

JURI AND HIS STUDENTS stayed on at Shisur for another month. They confirmed that people had dwelt here long before Ubar's hilltop fortress had been built, not just in the area of its spring but in the surrounding countryside as well. Juri spent considerable time mapping a satellite site he called "Flintknapper's Village." It was Neolithic — as old as 6000 B.C. — and he had a hunch that it had something to do with the beginnings of the People of 'Ad. He wanted to work out a sequence of settlement for the area, but that would have to be a project for the future, for it was getting hot on the edge of the Rub' al-Khali. By noon every day, the Arabian sun pushed temperatures well into the triple digits.

In early March, Juri and his students left Ubar to the watchful eyes of Baheet and Mabrook and retreated to the coast. There, almost immediately, he discovered that Ubar had a sister city: Ain Humran, a fortress overlooking the Arabian Sea. We had briefly visited Ain Humran during our reconnaissance back in 1989. It was a gloomy place on top of a gloomy hill, a heap of broken black masonry. "An old Portuguese fort," we had been told back then, a remnant of the swath cut through this part of the world by Afonso de Albuquerque in the 1480s. But now Juri saw that Ain Humran's walls and towers were much like Ubar's in width, height, arrow slots, and overall configuration. Like Ubar, the site controlled access to a nearby water source. Ain Humran, in fact, means "Red Springs."

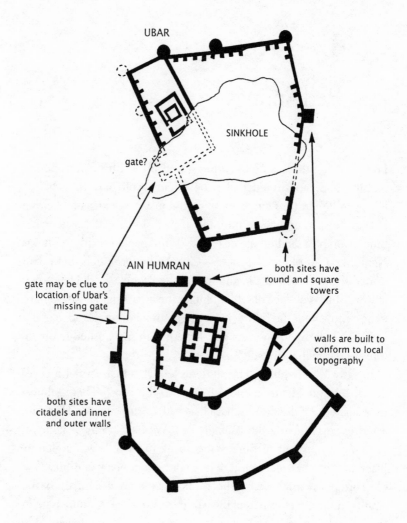

UBAR

SINKHOLE

gate?

AIN HUMRAN

both sites have round and square towers

gate may be clue to location of Ubar's missing gate

walls are built to conform to local topography

both sites have citadels and inner and outer walls

Ubar and Ain Humran: comparative site plans

The antiquity of the site was clinched as Juri and his students un-earthed shards of dot-and-circle pottery, the now recognizable hall-mark of the People of 'Ad. Here at last was a major coastal settlement built by the builders of Ubar.

With a city on the coast and another inland, the People of 'Ad could have dispatched their frankincense to faraway markets by either land or sea. Each route had its hazards. Ships were prey to storms and pirates; caravans were subject to high tolls and the depredations of brigands. Each year the 'Ad could have chosen the most promising path or could have used both routes. And if Ubar was "Omanum Emporium" on Ptolemy's map of Arabia, Ain Humran was most likely the same map's "Zaphar Metropolis," the "major city of Dho-far."

In late April, with hardly a riyal to spare, Juri and his students headed back to the United States. On the eve of his departure, the Sultanate of Oman promised to underwrite, for three years, the exca-vation of both Ubar and Ain Humran. In those years Juri would also record over 270 major and minor sites on the coast, in the Dhofar Mountains, and in the desert. They all had one thing in common: they defined in time and space the world of the ancient, no longer mythical, People of 'Ad.

18

Seasons in the Land of Frankincense

1993: Season two at Ubar and Ain Humran. At Ubar, the next year was the season of the Citadel.

Juri put his students and the Airwork volunteers to work clearing away the structure's rubble and sorting out its chronology. Its earliest phase dated to about 900 B.C., when it would have been the focus of a growing community that Juri called "Old Town." In approximately 350 B.C. it was enlarged, and walls and towers were added to create "New Town," in which the Citadel hovered over an enclosed market-place.

For the next six hundred years Ubar enjoyed the glory days of the incense trade. Its prosperity was mirrored in its finely crafted pottery — which, surprisingly, proved to be influenced more by the ancient cultures of the East than by those of the West. Though there was Greek and Roman ware here, much of Ubar's foreign (or foreign-inspired) pottery turned out to be Red Polished Indian Ware, a style indicative of a Mesopotamian/Persian influence.[1] In particular, Ubar seems to have had very strong ties to Parthia, a not very well known civilization that thrived between 400 B.C. and 300 A.D. and, in Juri's words, "gave the Romans fits" as their rivals to the east.

Descendants of nomadic horsemen, the Parthians were a prag-

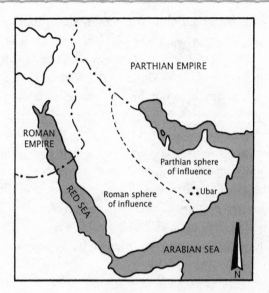

Foreign influences in Arabia, 200 B.C.

matic lot, with little interest in philosophy and the arts. They preferred baggy pants to flowing togas. They traded widely — from
China to Italy — and introduced to the West features of knighthood
and chivalry, including jousting and coats of arms. On the field of
battle, they frequently outflanked and outfoxed the Romans. The
Parthian cavalry would appear to flee, then its riders would swivel in
their saddles and with their bows deliver a "Parthian shot." The
expression is with us today when we refer to a parting shot.

Juri hypothesized that a Roman-Parthian division of the Mediterranean world and the adjoining Fertile Crescent extended to client
states in eastern versus western Arabia.[2]

As his team shoveled and sifted their way through the passageways
and chambers of Ubar's Citadel, Juri was pretty sure he was unearth-
ing an administrative center that could also have been a defensive

refuge in case of attack. But then there was the curious fact that the Citadel was built out of alignment with Ubar's nearby walls.

At the time, back in Los Angeles, I was intrigued by this anomaly and looked for parallels elsewhere in the Middle East, faxing Juri what I found. Some three hundred miles west of Ubar, the British archaeologist Gertrude Caton-Thompson had excavated the "Moon Temple" of Hureidha back in 1939. Noting that its corners were aligned with the cardinal directions, she theorized that its builders had been following the prevailing layout of Mesopotamian temples. Ubar's Citadel had precisely the same alignment (see below).

So it appeared that the Citadel was, in part at least, a temple or shrine as well as an administrative center.[3] The ancient Arabians saw no reason to keep their gods at arm's length from mammon. The gods

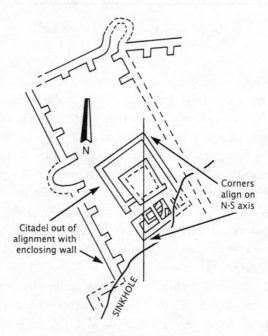

The Citadel at Ubar

blessed commerce, and commerce handsomely paid them back. The moneychangers weren't in the temple by accident; they were there by design.

As they excavated the Citadel, Juri's team cleared a passageway that led to a deep, finely plastered basin that once may have served a ritual function; perhaps it held water for ablutions, or possibly it was a place for the storage of temple frankincense. The passage turned to the left, then right, and ascended a well-constructed staircase — that ended in midair. Any chambers beyond — and the secrets they held — had sheared off and plunged into the site's sinkhole. The complete layout of the Citadel, unfortunately, would never be known.

Sometime after Ubar's great cataclysm, the Citadel and adjoining structures were partially reinhabited. In Tower #0 (so designated because it had been missed in the first season), Juri's team excavated 72 centimeters of stratigraphic deposits. (It was, he said, a "Sedimental Journey.") From 900 A.D. onward, the site had been rebuilt a half-dozen times, but with coarse mud brick rather than cut stone and fine plaster. As Juri commented, "Looks as if squatters moved in and hung out across from where the major collapse took place. Got by with as little as they could to make the place livable."

In early April of 1992, Juri shifted his crew to the coast, as he had the year before, and worked on through August. Kay and I joined him for a stint at Ain Humran. Kay, who had an eye for "surface collection," prowled the surrounding plain picking up bits and pieces of evidence of a sprawling agricultural community. I dug square #770 and didn't come up with an awful lot. When it came time for us to leave, I apologetically left the square to Juri's wife, Sandy, who along with their kids was in Oman for the summer. Volunteer Ian Brown also worked the square and, six days after we left, was startled to unearth a vessel flecked with purple paint and marked with six crosses. A Christian chalice!

This was a significant find, for where there was a chalice, there was

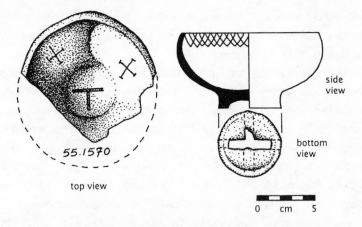

side
view

bottom
view

top view

0   cm   5

Chalice found at Ain Humran

almost certainly a church, perhaps a monastery. The chalice raised the possibility of monks sailing to this remote corner of Arabia and establishing an outpost of Christianity in an abandoned incense-trading center. Significantly, they would have been dwelling in the shadow of "the eastern mountain range" of Genesis, biblical mountains marking the edge of the known world.

There was once a considerable Christian presence in Arabia, and though documentation is sparse, there is a chance that Ain Humran was the missing "third church" founded by the Byzantine missionary Theophilus Indus in the middle 300s. The sites of Theophilus's first and second churches are known; the only clue to the locale of the third is that it was at a coastal emporium east of ancient Aden. It could well have been Ain Humran.

Theophilus, incidentally, was renowned for his DeMille-like miracles. A skeptical Arabian throng once challenged him: "Show us our Christ, alas!" The customary answer to this is an apologetic and gentle explanation that God works in mysterious ways and cannot be expected to work miracles on demand. Not in this case. Theophilus delivered. He looked skyward, "whereupon, after a terrible storm of thunder and lightning, Jesus Christ appeared in the air, surrounded with rays of glory, walking on a purple cloud, having a sword in his hand and an inestimable diadem on his head . . . [The challengers] were stricken blind, and recovered not till they were all baptized."[4]

At Ain Humran that second season, the Christian chalice was the sensational find. Juri also found and excavated the site's main gate, discovering that it had inner and outer pivoting doors, with a small chamber between. It intrigued him, for one feature he hadn't found at Ubar was a main gate, presumably because it had been destroyed when the site sank into the sands. The gate at Ain Humran gave him an idea of what to look for at Ubar the next season.

1994: *Season three at Ubar*. With the layout of Ain Humran in mind, it didn't take long for Juri to find a matching gate at Ubar. It was in the western wall, between Tower #5 and the Citadel. At least it once was. Only the outer doorjambs remained; the rest had collapsed into the sinkhole. In search of the missing gate, Juri and his students sank a three-meter-square shaft in the sinkhole's sands.

Digging back in time, they first made their way through sand mixed with animal droppings and bits of bedouin bowls. For the last fifteen hundred–plus years, nomadic bedouin had watered their camels and goats here. Otherwise, there were no signs of occupation. Then, from one and a half meters on down, Juri's team unearthed, one by one, stones that in cut and dimension precisely matched the masonry of what remained of the gate above.

Below the masonry of the crumbled and fallen gate were fragments of jagged, raw rock, once the gate's bedrock footing. Farther down, Juri and his students sifted through sand containing fragments of ancient pottery and bits of flint, evidence of the site's long occupation before its collapse. In the stratigraphy of the sinkhole's sands, Juri proved beyond a doubt that a single violent cataclysm had led to

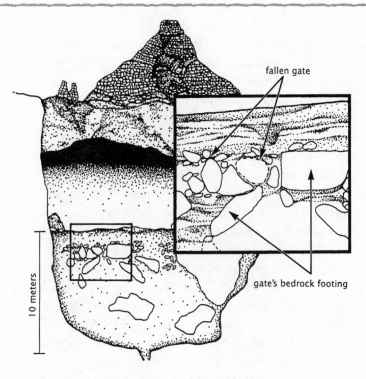

fallen gate

10 meters

gate's bedrock footing

Cross-section of the sinkhole

Ubar's abandonment. The sands below the fallen gate contained evidence of long and meaningful settlement. The sands above the gate contained next to nothing.

Unfortunately, further digging would almost certainly destabilize the sinkhole and precipitate a disastrous cave-in, and so it is that a significant portion of the fallen Citadel would remain buried.

Season three was to be the last at Ubar. The city would still have its secrets and enduring mysteries. Beneath the rubble, beneath the sands, there still could be — who knows? — inscriptions, idols, skeletons, even treasure. Which is, perhaps, as it should be.

1995: Season four, the last in the land of frankincense. Juri and his crew devoted their final season to a wide-ranging search for evidence of the presence of the People of 'Ad in the Dhofar Mountains and on the shores of the Arabian Sea. His archaeological sequence for the region now lacked but a single horizon: the Bronze Age, which in that part of the world was 2350–1200 B.C. It was Airwork volunteer Sean Bowler who at Taqa, on the coast, found the first tiny evidence of that era: a single bronze fishhook. And it was student Jim Brake who hiked up a hill into a Bronze Age bonanza.

The main road from the coast up into the Dhofar Mountains passed a solitary spreading olive tree, unique in the surrounding broken-limestone landscape. A left turn at the tree led to a high desert valley where frankincense groves were still to be found here and there. We had driven this road dozens of times. We had stopped to examine and photograph the trees; we had watched as tribesmen had slashed their branches and harvested frankincense crystals. This final summer, something caught Juri's eye. As usual, it was a rock, this one broken into three pieces. It was a monolith, a fallen pillar. Originally three meters high, it appeared to be funereal in nature, for it marked a burial site.

As Juri measured and photographed the pillar (and estimated its weight at five or more tons), student Jim Brake crossed the road and climbed to the crest of a low hill, only to come running back down. There were ruins up there; they went from hill to hill to hill. Jim had happened on, in Juri's words, "a monster Bronze Age site."

Set on the banks of what were once three converging streams, site Hagif #240 rambled on for a good three miles. Judging from their foundations, the village's houses had been impressive, with entries flanked by rows of massive standing stones. Hagif not only proved to be the largest Bronze Age site in all Oman, it filled in a key period in the story of the People of 'Ad. Around 2500 B.C., the rains that had long blessed southern Arabia withdrew, initiating an arid period that continues to the present. The area in which frankincense flourished likely shrank to its current range on the back slope of the Dhofar Mountains. The majority of the region's seminomadic dwellers followed this retreat and settled in the heart of the remaining frankincense groves. Hagif would have been their principal site.

At the same time, far out in what had become a waterless, hostile desert, a hardy minority of these people withdrew to a last freely running spring, flowing from the cavern of Shisur. They would become our Ubarites. As we shall see, drought and desertification worked to their advantage and enriched them, for their modest settlement now became the *only* viable water and rest stop for caravans carrying incense across the sands of the Rub' al-Khali.

Time and time again in the last few years, Juri Zarins had shaken his head and told us how experts often dismissed an area as unimportant because, truth be told, they hadn't spent enough time there to have a really good look around. This had certainly proved the case with the mountains and interior of Dhofar. Well into the 1980s, it was believed that the true land of frankincense was to the west, in the kingdom of the Hadramaut. Juri and his crew had proved otherwise and had done it so thoroughly that it was now possible to tentatively reconstruct the history and life of the once mythical People of 'Ad.

The story told in the next several chapters is a story framed by archaeological evidence, including the results of carbon-14 dating, and filled in with material from classical accounts. From time to time, it incorporates traditions of desert life that have survived intact into our century.[5]

Setting foot in Oman five years ago, we saw Ubar and the People of 'Ad "through a glass, darkly; but then face to face."

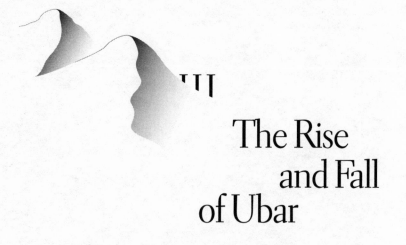

III

The Rise
and Fall
of Ubar

19

Older Than 'Ad

IN THE VOCABULARY of our bedouin friends, "old" meant when their grandfathers were alive, and "really old" meant a hundred or so years ago. If you were interested in something thousands of years old, you said "as old as 'Ad" or "older than 'Ad." In his desert archaeology, Juris Zarins was interested not only in Ubar's classical period — its rise and citification — but in times "older than 'Ad." Long interested in the origins of things, he sought the very first people to walk the surrounding landscape. It wouldn't be easy, for over the millennia the desert's geology had not so subtly shifted, hiding older artifacts.

Deposition and erosion had conspired to change the sites of ancient, logically situated camps into places where now nobody in his right mind would choose to stay. Prowling the desert, Juri would ask himself, "If I were early man, where would I camp?" Depending on where he was, the answer might be "Where I can see game" or "Right by the river" (for there once had been rivers and lakes out here). Often referring to space images, he would then work out how such a site might have been affected by the region's evolving geology.

Space imaging is what one day drew him sixty miles south of Shisur to the banks of the dry Wadi Ghadun. The wadi was deep, cut into the desert by infrequent but torrential floods and, a very long time ago, by a slow-moving river. Five terraces now stairstepped up its banks. The top terrace, Juri knew, would be the oldest. Early man

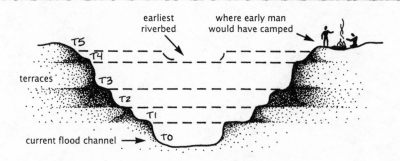

Cross-section of Wadi Ghadun

would have camped there first, then gradually moved down to stay close to water as the wadi eroded and deepened. Walking the top terrace, Juri frequently stooped down to pick up a stone or two, only to pronounce them AFRs — worthless — and throw them over his shoulder. But then he found a concentration of stones that fit comfortably in the palm of his hand. His notes matter-of-factly record: "The ferruginous quartzite specimens are very windworn, but consist of flakes, choppers and some scrapers. Typologically, they represent the oldest site yet found in Dhofar."

What Juri had found was a sampling of Acheulean utensils that were 700,000 years (or more) old, the handiwork of long-vanished *Homo erectus.* Here was the beginning, Chapter One of the Ubar story. With skill, tenacity, and luck, Juri was subsequently able to detect the footsteps — all at or near Shisur — of man's journey from that time to the present.

For the better part of a million years our distant ancestor *Homo erectus,* upright but not very bright, roamed Arabia. Then, approximately one hundred thousand years ago, *Homo erectus* was displaced, as our direct ancestor, *Homo sapiens,* migrated out of Africa and across Arabia. It was not a difficult journey, for there was then a land bridge at the south end of the Red Sea, and Arabia at that time was verdant and welcoming. Every year life-giving monsoon rains swept across the peninsula. The rains gave birth to rivers and created a thousand or more lakes, home to water buffalo and hippopotamuses. (In the sands of ancient lakebeds geologists have found intact fossilized hippo teeth, so well preserved they could have been lost just yesterday.) Clouds of dark smoke rose from the shores of these lakes, from fires set by *Homo sapiens* to flush out wild cattle, goats, oryxes, gazelles, and possibly camels and hartebeests. The game was roasted at camps on ridges and hilltops all around Shisur. At these sites, early man had open-air workshops for manufacturing the huge blades he favored for his spears.[1]

But then, some twenty thousand years ago, the rains withdrew.[2] The rivers and lakes of the Rub' al-Khali dried up, and violent winds tore at their sandy beds and reworked them into vast fields of dunes. The birds fled, leaving the sky to a merciless sun. Daily temperatures soared to over 130 degrees in the shade, if there was any shade. Early man cleared out, in all probability retreating to the north and the land of the Fertile Crescent.[3]

For the next hundred centuries there was no appreciable rainfall, and in the whole peninsula not a trace of human occupation. All that survived, in isolated pockets, were highly drought-tolerant animals and plants, among them a small, scraggly tree that favored a harsh limestone substrate and warded off other vegetation with toxic terpenes spread from its roots: the frankincense tree.

About eight to ten thousand years ago the rains returned to Arabia, and wanderers from the north appeared on the peninsula's empty stage. They came from what has been called the "proto-Semitic homeland," an arc stretching from northern Egypt up into Syria. In an amazingly short time — as little as two hundred years — they repopulated all of Arabia. The pride and sustenance of these people was their cattle; their progress through the peninsula is marked by images of cattle they pecked on blackened rocks. At their campsites, these new pastoral nomads gazed skyward and imagined the stars as the cattle of the moon, penned only by the far horizons.

By the time they reached the Dhofar Mountains (the only place in Arabia where a cattle culture still survives), a group of these wanderers had most likely achieved a tribal identity, an identity that would become the People of 'Ad. They settled down and enjoyed the favors of a land that every year was becoming greener and more bountiful. The monsoon rains spilled over the mountains and watered the land beyond. The desert bloomed, soaked up the rains, and issued them forth as springs. Bubbling up through an ancient cavern, one spring would someday be called Shisur, the spring of "the cleft."

The early People of 'Ad camped near the spring but probably not at it; because animals came to drink there, Shisur was an ideal place to trap game. Shisur's Neolithic game trap was cleverly laid out.

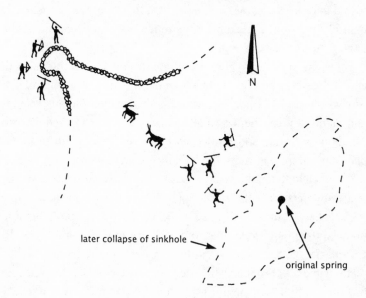

Neolithic animal trap at Shisur

When gazelles and oryxes came to drink, beaters would approach from the east and noisily drive them between two rock walls to the west of the spring. The narrowing walls forced the panicked, confused animals into a rock circle. There, waiting hunters would rise up to take their prey with arrows, spears, and nets.[4]

At nearby sites, such as Flintknapper's Village, the People of 'Ad would have enjoyed a good life, as good as the late Stone Age allowed. Goats had been domesticated, and their long hair was loomed to create spacious, comfortable tents. Domesticated cattle provided skins, milk, and meat. Though game wasn't as plentiful as it had been when they first settled here, the People of 'Ad sharpened their hunting skills by crafting finer, more effective arrowheads.

At first the 'Adites worked just one side of their large flint arrowheads, but then, influenced by samples imported from the north, they worked both sides and added a barb. Finally, in an advance that was their own invention, the arrowheads of 'Ad were streamlined and deftly serrated, with a ridge running down the middle. They were contoured "trihedral rods" (as classified by archaeologists).

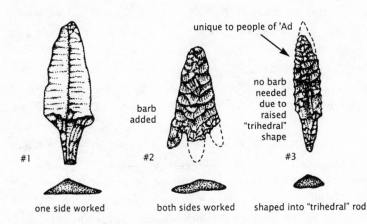

Evolution of arrowhead technology

This shape could be achieved only with skilled, twisting blows of stone on stone. Found throughout Dhofar, these trihedral rods defined, at an early stage in their existence, the range of the People of 'Ad. Remarkably, it was a territory that would be theirs for the next 5,000 years (circa 4500 B.C. to 500 A.D.).

The life of the early 'Adites centered on their campfires. It was there that they crafted their arrowheads and stone tools, and it was there, quite by accident, that they may have discovered the fragrance and uses of frankincense. Imagine an extended family camped at the same place for several months. With the supply of deadfall firewood exhausted, a couple of children might have been given a hand axe and asked to cut an armload of branches from a nearby scraggly tree, no taller than they. As the fire was kindled, an unusually white smoke curled up, and instead of watery eyes and coughs, it prompted appreciative sniffs and sighs. The smoke of the frankincense was sweet and clean. If in their early belief system the People of 'Ad had notions of Paradise, its scent was that of frankincense.

The 'Adites doubtless found many uses for frankincense. What better offering for their animistic and celestial deities? It had practical applications as well. It took the edge off the smell of well-worn garments; it sweetened drinking water; it hastened the healing of wounds. After dark an 'Adite could shape a blade by the intense, almost supernatural light of burning frankincense. Quite naturally, word of this wonderful substance spread, and samples were traded to nearby tribes and eventually to the civilizations of distant lands.

As early as 5000 B.C., there is indirect evidence that the northern Mesopotamian city of Ubaid imported pearls, precious stones, and incense from Arabia. The Ubaid culture was in time supplanted by the civilization of the Sumerians, and at their great city of Uruk several bas-reliefs illustrate offerings of incense to the sun god and his consort. Researching Sumerian cuneiform tablets, Juris Zarins found that their deities were in the earliest years purified with the burning of cedar brought from Lebanon. But then, according to a text dated to 2350 B.C., these deities were offered incense:

(SHIM = incense)

More specifically, the gods were offered what was probably frankincense:

(SHIM.GIG = frankincense)

The displacement of cedar by frankincense as a temple offering would have been expedited by the domestication of a sure-footed, tough beast of burden — the donkey. The first long-range caravans were donkey caravans, and with them a "merchant of aromatics" could have made yearly forays into the heart of Arabia.

(GARASH.SHIM = merchant of aromatics)

The ritual burning of frankincense became a call to the gods who, as it is recounted in the epic of *Gilgamesh*, "smelled the sweet savor. The gods gathered like flies over the sacrificer." Though the Sumerians didn't have a very high opinion ("like flies") of their gods, smoke curling heavenward conveyed pleas, expressed gratitude, offered atonement. Indeed, the German incense historian Walter Müller believes that throughout the Middle East, frankincense was considered to have unusual expiatory power, much like the power gained by sacrificing an animal, for "its resin was considered to be the blood of a tree, which was taken to be animate and divine."[5]

It is in frankincense that the subsequent story of the rise and fall of the People of 'Ad would be written. The resin became an integral part of their lives over the next several millennia. Frankincense would beguile them. It would cause them to prosper and take on the airs of a classic civilization. And then, at least according to legend, the 'Ad became arrogant and unjust and were punished. Their desert city of Ubar was destroyed. At the same time, because of the rise of Christianity, the demand for frankincense fell off. And, whether or not they deserved God's wrath, the 'Ad would be left illiterate and poor, dwellers in the ruins of their past glory.

20

The Incense Trade

ONE REASON the People of 'Ad were so long cloaked in mystery is that outsiders were not welcome in their land; the harvesting of frankincense was a secretive affair. Nevertheless, Pliny the Elder managed to come by a good description of the process. Considerable pains were taken not to injure the trees, and timing was important. Only under the best conditions would the trees produce the finest, most fragrant incense. A midsummer harvest was augured by

> . . . the rising of the Dog Star, a period when the heat is most intense; on which occasion they cut the tree where the bark appears to be the fullest of juice, and extremely thin, from being distended to the greatest extent. The incision thus made is gradually extended, but nothing is removed; the consequence of which is, that an unctuous foam oozes forth, which gradually coagulates and thickens. . . this juice is received upon mats of palm-leaves. . . The incense which has accumulated during the summer is gathered in the autumn: it is the purest of all, and is of a white color.[1]

Around 2000 B.C., the range of the frankincense tree, encouraged by abundant rains, may well have extended out to and even beyond the spring at Shisur. In any case, the surrounding oasis was an ideal staging area for caravans heading north. It provided fodder for donkeys and dates for their drivers. Its palms shaded a primitive market, a place to barter for pack saddles, sacks of salt, obsidian tools, and luxuries like beads and decorative shells.[2]

Forming at Shisur and striking north, the donkey caravans followed a trail of springs and seasonal lakes spaced no more than a day apart. Over the years, over the centuries, the footprints of thousands and thousands of animals compacted the desert pavement, creating a singular track, which at a later time would be called "the road to Ubar."

This great track led to Jabrin, a Mesopotamian-controlled oasis on the far northern edge of the Rub' al-Khali, where in all probability the People of 'Ad sold (or bartered) their frankincense, then hastened home. Even though the climate was cooler than it is now and moist, the round trip was arduous. If a caravan failed to reach a spring or found it fouled or dry, animals and men could perish.

As they crossed the Rub' al-Khali, early caravans would have sighted a wild, humpbacked, ungainly animal that was at home in the desert wasteland. According to legend, the 'Adites thought it a creature conjured by djinns. But it would transform their lives. Over the previous centuries, man's livelihood in Arabia had benefited from the domestication of a sequence of animals: cattle provided mobility and sustenance, goats offered tents and textiles (as well as sustenance), and donkeys opened the way for long-range travel and trade. Finally, the camel was to make distant trade commercially viable year in and year out, in bountiful times and bad.

The camel could carry a six-hundred-pound load and could go two weeks or more without water. No longer was it necessary for caravans to meander from spring to spring; they could now move in straight lines. On level ground a camel caravan could cover thirty miles in a day. If need be, a well-bred camel could cover close to two hundred miles in a twenty-four-hour period.

Legend — and some evidence — has it that the camel was first domesticated in southern Arabia by the People of 'Ad. There, more than fifty "houses" (breeds) of camels were developed. The animals were cranky, had bad breath, and were certainly not beautiful in any conventional sense of the word. Yet their owners could gaze into a camel's great eyes and see both economic gain and a soulmate. In an age-old ditty of the desert, the beauty — and worth — of a woman is measured against the Banat Safar, a "house" of camels:

The fairness of beautiful girls
Is that of the Banat Safar.
Sa'id's daughter approaching a campfire
Is like a camel descending a difficult pass
             [i.e.: its head, like the girl's, turns superciliously from
             side to side]
Her fresh face is like a camel's flesh
Which the dew has not struck, nor the cold.[3]

With the domestication of the camel, the pace of the incense trade quickened. Caravans could cross the Rub' al-Khali in less than a month, and their frankincense was then carried either north to Mesopotamia or west to the Red Sea. There it was loaded on boats bound for Egypt, where it was greatly valued as early as 2800 B.C. As an offering worthy of "the great company of the gods," the Egyptian *Book of the Dead* considered incense far more than a ceremonial trapping: the incense itself was holy. At a funeral, the text instructed: "Thou shalt cast incense into the fire on *behalf* of Osiris" (rather than offer it *to* Osiris). Frankincense enhanced the afterlife journey of the deceased. In the words of the ritual *Pyramid Text*, "A stairway to the sky is set up for me that I may ascend on it to the sky, and I ascend on the smoke of the great censing."[4]

As they learned of the holiness accorded frankincense, the People of 'Ad may have ritualized its gathering. Pliny tells us that the individuals selected to harvest it were "called sacred, and . . . not allowed, while pruning the trees or gathering the harvest, to receive any pollution, either by intercourse with women, or coming in contact with the dead; in this way the price of the commodity is increased owing to the scruples of religion."[5]

Though the religious (versus economic) ardor of the 'Ad may be a little suspect in this case, they almost certainly would have had a belief system drawing on both their Semitic heritage and practices in Mesopotamia observed in the course of their incense trade. When times were good, as they had been for as long as anyone could remember, the 'Ad probably took their gods for granted, paying them minimal heed.

Then, sometime around 2500 B.C., bad times came to the People of 'Ad. The rains ceased to fall at Shisur. First one year, then another, then every year the monsoon rains failed to crest the Dhofar Mountains and water the land beyond. At Shisur the 'Ad would have, in all earnestness now, turned to a rudimentary temple. It may have been a sacred tent, a brush-walled enclosure, or perhaps just a low wall of uncut stones outlining a *haram*, a sacred space. In its precinct frankincense would have been placed in small burners — miniature fire altars — and offered to an uncut stone, a betyl, which was the dwelling place of Sada, the bringer of rain.

The slow-moving river near Shisur dried up, and any frankincense trees around Shisur died off. The god Sada failed the 'Ad. It was then that in large numbers the People of 'Ad withdrew to the south and settled in the highlands of the Dhofar Mountains. There frankincense still flourished, and there was water. Where three springs flowed from a low ridge, the Bronze Age mountain town of Hagif arose. Considerable effort went into its domed houses built of branches set in megalithic foundations. There was a sense of permanence here, new to southern Arabia. As one generation gave way to the next, new houses were built and older ones converted to impressive graves; Hagif became a rambling assembly of the dead and the living, sprawling three miles across the incense land.

In the desert beyond, all was not abandoned. At this time a small clan of the People of 'Ad achieved a new importance. The prototypical people of Ubar controlled the oasis and spring at Shisur, now the only place where fresh water flowed. In thirsty Arabia, water became a major economic determinant. Caravans could still venture across the Rub' al-Khali, but not without stopping at Shisur to rest and — for the appropriate tribute — to water their camels.

Though Shisur may still have been the modest place it had been since the discovery of frankincense, now it was the *only* settlement of consequence beyond the Dhofar Mountains. In contrast to the 'Adites living in the mountains, the Ubarites dwelt not in stone and brush houses but in tents. Compared to the cramped, smelly structures at Hagif in the mountains, tents were spacious and airy. Their flaps could be adjusted to bar a sandstorm or to capture a gentle breeze. They could be relocated when garbage accumulated. When in summer the desert heat became truly oppressive, the Ubarites could fold their tents and retreat to the higher elevations of the Dhofar Mountains, to live with their 'Adite kin and help with the frankincense harvest.

For several hundred years, Arabia baked and dozed in the sun, "fiery hot and scorched," in the words of Strabo. The incense trade continued with few innovations, few ups and downs, few threats. Then, between 1400 and 900 B.C., the picture changed. To the east of Shisur, toward the Red Sea, four new kingdoms rose and prospered: Ma'in, Saba (or Sheba), Qataban, and Hadramaut. These city-states raised dams, dug irrigation channels, and developed an impressive agriculture. They also sought a share of the incense trade. They harvested some aromatics themselves, including myrrh and low-grade frankincense. But for the finest frankincense, they were reliant on the People of 'Ad.

In about 950 B.C., the Bible tells us, a queen of the Sabaeans journeyed north to Jerusalem to the court of King Solomon to work out a trade agreement to supply incense to Israel and other countries of the eastern Mediterranean. (The Israelites would serve as brokers.) Solomon's knowledge of the precise origin of the finest frankincense was probably hazy, and the queen was not about to set him straight.

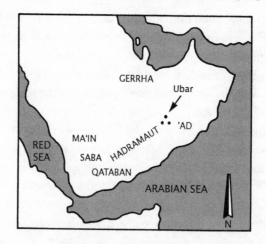

Kingdoms in southern Arabia, 350 B.C.

She professed to be dazzled by his wisdom and, deal done, went on her way.

Incense had become indispensable in the Israelites' rituals. Even as a wanderer in the wilderness, Moses told his people to "take every man his censer, and put incense in them" (Numbers 16:17). Later, in the face of a plague, he told his brother Aaron, the high priest, to "take his censer and carry it quickly among the congregation, and make atonement for them. . . . And he stood between the dead and the living; and the plague was stayed" (Numbers 16:46–48). In the Great Temple of Jerusalem, frankincense alone among incenses was reserved for the worship of Yahweh; any misuse or profanation of it was punishable by death.

At Shisur, the needs of Israel and the other civilizations of the eastern Mediterranean had their impact, as did the uncomfortable proximity of Arabia's new city-states, especially the kingdom of the Hadramaut. There was an impetus for the Ubarites to assert them-selves in trade — and in their standing with the unseen, with their gods.

Sometime after 1000 B.C., a cluster of stone, mud, and brush build-ings rose on the hilltop overlooking Shisur's spring. Juris Zarins dubbed this "Old Town." At its heart was a small temple. This modest complex to some extent echoed the faraway, highly developed relig-ion-centered architecture of Mesopotamia. The mythologist Joseph Campbell has made much of the Mesopotamian "hieratic city." Rather extravagantly, he declares:

> The whole city now is conceived as an imitation on earth of the celestial order — a sociological middle cosmos, or mesocosm, be-tween the macrocosm of the universe and the microcosm of the individual, making visible their essential form. . . [It is] the sanctu-ary of the temple, where the earthly and heavenly powers join. The four sides of the temple tower, oriented to the four points of the compass, come together at this fifth point, where the energy of the pleroma enters time. . . . And this mesocosm is the entire context of the body social, which is thus a kind of living poem, hymn, or icon, of mud and reeds, and of flesh and blood, and of dreams, fashioned into the art form of the hieratic city state.[6]

How much of this the Mesopotamians actually had in mind is uncertain, and how much of it rubbed off on the Ubarites is even more questionable. Yet it was from Mesopotamia that the hieratic design of cities disseminated, and that design dovetailed with a belief, common to kings and commoners, that the gods dwelt in a celestial realm that was more splendid, more ordered than life in the dust of the earth. At one time man may even have enjoyed that life, but no more; he had long since sullied the earth's primordial harmony.

In temple and city building, early civilizations sought to re-create or at least mirror a celestial realm. To the Israelites, the Mesopota-mians, and the Ubarites alike, there was no question that life was fragile, shadowed by mortality. Yet if men built houses — temples — as their gods built houses, couldn't they then share their enviable power?

As legend has it, Ubar became "an imitation of Paradise." And so it may have been. Its verdant oasis would have been a striking contrast to the surrounding parched, dead, and dying landscape. The focus of that oasis was now not only a life-giving spring but a temple that, however small and simple, was oriented to the cardinal directions (as Campbell mentioned) and looked upward to the heavens and eternity.

What gods the Ubarites saw as ascendant in the heavens is unclear. A rain or storm god called Sada appears to have had some importance, but he may have fallen into disfavor, considering how little he'd done in centuries of merciless drought. By the same token, the Ubarites would have had quite enough of the sun god favored in the Fertile Crescent (where that god brought warmth and fertility rather than heat and death). A moon god was almost certainly preeminent, his strength and power symbolized by the crescent-horned bull. But it is doubtful that the Ubarites worshipped the bull as a graven image; they would have been content to burn frankincense before an elemental, uncut rock. Or they would, in the coolness of the night, have looked skyward as the moon itself rose and dominated the night sky. The phases of the moon were carefully observed, and a phase of the quarter moon — Il or Ilah — came over the centuries to be taken as a general term for God; it is the root of the Hebrew El or Elohim and of the Arabic Allah.

With their first temple, the Ubarites staked their place in the cosmological world and in the temporal world as well. The temple served several down-to-earth purposes. On its roof, lookouts could stand watch over a 360-degree view of the surrounding desert. As in the biblical Song of Songs (3:6), the Ubarites could look to the far horizon and ask, "What is this coming up from the desert like a column of smoke, breathing of myrrh and frankincense?" Most often it would be friendly caravans. Though marauders were an occasional problem, they would think twice before attacking Ubar. If they as much as tried to water their animals at the Shisur spring, they would be met by a hail of arrows from the stone structures and temple above.

With their temple, the Ubarites were both reaching for the heavens and protecting their flanks. The temple compound would have served as a lockup, a safe deposit for whatever gold or treasure the Ubarites possessed. Even better, it became a means of acquiring wealth, for the temple's existence justified the demand for tolls on passing caravans. The Ubarites had probably long benefited from the needs of passing caravans. Now the gods needed to be fed, and what did they appreciate more than frankincense, the very "food of the gods"? Judging from accounts of similar caravansaries, the toll demanded (subject, as always, to negotiation) could have been a tenth of a caravan's cargo.

At the time Ubar's Old Town was built, a distinctive written alphabet appeared in southern Arabia. It was used by the newly risen kingdoms of Ma'in, Saba, Qataban, and Hadramaut. But in the land of the People of 'Ad this alphabet was expanded; eight more letters were needed to record the more complex language spoken there. The language of the 'Adites, then, may well be Arabia's oldest, for it is a general rule of linguistics that languages simplify — and lose sounds and letters — as they evolve.[7]

The origin of the 'Adites' ancient language is reflected in its construction. It is clearly an early Semitic language, yet it appears to incorporate Mesopotamian-inspired verbs and word endings. Such were the People of 'Ad: a Semitic group with long-standing ties to Mesopotamia. Those ties, though, may not have always been friendly. A Mesopotamian inscription from circa 720 B.C. tells us that the armies of Tiglath-pileser III marched down the incense trade route and pushed back the Arabians. The Arabians in question were probably not the 'Adites themselves, but their trading partners to the north, the Gerrhans. Originally a pack of shepherd-brigands exiled from Babylon, the Gerrhans not only overran the oasis at Jabrin (directly across the Rub' al-Khali from Ubar), but piratically infested the waters of the Persian Gulf. In 694 B.C., Sennacherib put them down and, for a time, confined them to their city of Gerrha. But they were soon out roaming the desert and opening up caravan routes across northern Arabia.

The Gerrhans and the 'Ad, one suspects, were cut from similar cloth. Accepting the lot dealt them by life in the desert, they were ardent and not altogether trustworthy traders. They were intolerant of restraint, a thorn in the side of classical civilizations, whose wealth they considered their godsend.

Classical authorities tell us that for several centuries, the Gerrhans transshipped the frankincense of the People of 'Ad to both Mesopotamia and the Mediterranean. The demands of the Greeks, then the Romans, became prodigious. In the early era described in Homer's *Odyssey*, the Greeks made do "with the fragrance of flaming cedarwood logs and straight-grained incense trees." When frankincense made its way up from Arabia, they were enthralled. The goddess Aphrodite was especially appreciative of frankincense, as was Alexander the Great, the first ruler known to have it burned in his honor.

As well as an offering to the gods, the Romans used frankincense as a balm to the body in this life and a means of easing it into the next. It was an ingredient in cosmetics and perfumes. It was materia medica, good for "broken heads . . . and to bind bloody wounds and assuage malignant ulcers about the seat."[8] The Romans' most copious use of frankincense was in their rites of cremation, for it pleased the gods, and, as down-to-earth Pliny notes, disguised the odor of burning bodies. He tells us that an entire year's production of Arabian frankincense was heaped upon the funeral pyre of Poppaea Sabina, Nero's wife. Pliny sums it up: "It is the luxury of man, which is displayed even in the paraphernalia of death, that has rendered Arabia thus happy."[9]

How it must have galled the Greeks and Romans that distant foreigners — smug in their Arabia Felix — profited so from their profligate lifestyle. Though parts of the peninsula were tenuously within their sphere of influence, the Greeks and Romans knew little — heard rumors, at best — of the secret city that was a key source of their frankincense. It was an uncertain dot on Claudius Ptolemy's *Sexta Tabula Asiae*, hidden in the desert of the Iobaritae, the Ubarites.

It is now that the myth of Ubar comes into play and runs a course parallel to the site's archaeological record. Archaeology is history; myth is imaginative history. The two should probably not be mixed, but with Ubar, the temptation is hard to resist. Be patient then, with the next chapter, which speculates — very conjecturally — what might have happened at Ubar in a season of its glory. It is written with the caveat oft used by Arab historians and storytellers: "But God, however, knows best."

21

Khuljan's City

IN THE LAND OF 'AD, in Arabia Felix, a moonlit late summer's night in 350 B.C.[1]

Roped four abreast, the column of camels shuffled across the desert plain. Their drivers intermittently dozed in the saddle, then jerked awake, often just in time to keep from pitching to the ground as a camel stumbled, which they occasionally did when moving by night. An old man, "as old as 'Ad," his companions joked, cleared his throat and chanted . . .

> After the sun has set, in the watches of the night,
> May the god of the moon whiten our faces . . .[2]

The first caravan of the year, it was larger than usual, for its camels were laden not only with frankincense but with bags of rock salt. The salt was for a contingent of stonemasons and their apprentices, who were following on foot. The time had come to enlarge Ubar's temple and enclose and fortify its spring. Six days had passed since the caravan left the Dhofar Mountains. The little water left in the goatskins was fetid, barely drinkable.

Off to the east, the sky lightened. When the sun rose this day they didn't stop to rest but kept on. In protest, the camels pitifully gurgled, then brayed and balked. Their drivers were tempted to strike them but didn't, for it would damage their qualities. Instead they shouted, "Evil pestilence upon you!" and "Come to you death!" In their hearts they meant nothing of the kind, for their camels were their life.

It was a sharp-eyed boy who first spotted a tiny smudge of green on the horizon, appearing, then vanishing in a mirage.

"Hai! Our deliverance. Ubar," breathed the old man, as he had for most of his forty years.

Drawing closer and wending up the low hill surmounted by Ubar's old temple, the caravan was met by the small garrison — no more than a dozen men — that stood watch over the site during the hottest months. The camel drivers unloaded their cargo of salt and frankincense. After resting for a few days, they would return to the mountains for another load. The stonemasons unpacked their sledges and chisels and examined the site. They hammered at the limestone rising from the sand to the west of the temple. It was of adequate quality, and quarrying it would create a dry moat, a bonus to the fortification of Ubar. The stonemasons splashed in the water of the settlement's clear, cool spring, had their fill of fresh dates, then slept through the afternoon and night.

Up well before dawn the next day, the masons saw to the digging of shallow trenches to the south of the temple. They lined the trenches with goat droppings, then packed in a layer of rock salt. By midafternoon they were able to step back and view the footings of an inner and outer gate. As protection against djinns, they drove spikes into the ground at the gate's four corners.

From the shadows of the nearby temple a *kahin*, a soothsayer-priest, came forth. He rapped a hide drum with his knuckles, slowly at first. Two girls stepped forth from the temple, dancing. The older led and the younger followed, imitating, as best she could, her partner's spontaneous movements. They danced upon the salt. The kahin beat his drum faster, as fast as he could. The dancers' stampings and gyrations were frenzied now. Camel drivers drifted over from their camp and joined the masons in clapping to the rhythm of the dance, accented now by the bleating of a tethered goat.

With a cry of "For the face of the lord of the moon!" the kahin unsheathed a dagger and slashed the goat's throat. He lifted the animal and carried it about so that its blood would fall on the gate's four corners. The girl dancers pressed their hands in the blood, and raised them high to the accompaniment of wild ululations. For good measure, the kahin carried the dying, bleeding goat down the hillside, where more trenches would be dug and walls and towers would rise. That evening, the meat of the goat was divided, one measure to the kahin and four to the masons, builders of a new Ubar.

In the heat and dust of the next few weeks, the gate was fitted with heavy wooden doors and completed. At the same time, the masons and their apprentices were at work on Ubar's walls. They laid three to five courses of cut blocks, filling gaps by wedging in small stones selected from the many scattered around the spring (many of these were Neolithic tools, which would now be preserved in Ubar's walls). Every few paces along the walls, they fashioned angled arrow slits. The masons kept the width of the walls to a consistent elbow-to-fingertip span (92 centimeters). Unlike other sites in the Middle East, where imposing mud-brick walls would be built on this stonework, at Ubar the foundations were surmounted by a *duwwar*, a fence woven of gnarled branches and brush.[3]

At the corners of Ubar's rising fortress, the masons erected sturdy towers. Additional towers guarded vulnerable points in the wall.

To complete their work, the masons may have returned the next season and even the one after that. The fortress they built served the essential needs of an early city: exchange and defense. With its welcoming gate and large interior court and spring, Ubar could accommodate the ebb and flow of desert trade. In the event of an attack, that same gate could swing shut and secure the Ubarites, their animals, and their frankincense.

At some point in the course of the construction, a lookout on duty in one of the towers would have caught sight of a column of dust rapidly approaching: the king of the People of 'Ad and his retinue. To lend him a name, call him, as in myth, King Khuljan ibn al-Dahn ibn 'Ad.

Khuljan would have cut quite a figure. Riding a sleek stallion, he wore a scarlet robe fastened over his left shoulder. In preparation for his arrival at Ubar, the royal barber had woven the king's long hair in plaits and dyed it blue with the juice of the *nil* plant and had blued his face as well.

Khuljan wore no crown but rather a five-thonged leather head-band wrapped with bands of gold and silver. On arriving at Ubar's new gate, he would have been honored with clouds of incense, and commoners falling to their knees to kiss his knees. Those of higher station who knew him well kissed his wrists, elbows, and shoulders. Little boys, if they dared, jumped to clap their hands beneath the king's nostrils, so they might acquire the virtue of his breath. The kahin who looked after the temple offered Khuljan hen's eggs, which the king dashed on the outer and inner thresholds of the gate, dedicating it to the glory of the gods and to the prosperity of his tribe. After inspecting Ubar's fortifications and enlarged temple, the king retired to his tent. It had, as always, been a long journey.

In the manner of Arabian kings, Khuljan would have held morning court (or majlis, as it later came to be called) in the shade of the gate. At times it was a court of judgment, with the king acting as "master of ordeals." As an accused man was brought forth, Khuljan drew his finely wrought bronze dagger and laid it on the coals of the fire kept by the soldiers on duty at the gate. He chatted with them as they watched it brighten and redden; it would swiftly determine whether the accused was "of gold or of iron," an innocent or a scoundrel. Turning to him, Khuljan ordered that he open his mouth and show his tongue. The king then took the tip of the man's tongue between finger and thumb with one hand, and with the other raised his dagger to his own lips and, almost kissing it, whispered, "O fire, O fire, be cold and at peace."

Swiftly and without hesitation, Khuljan pressed the flat of the dagger upon the man's outstretched tongue, turned it over, and pressed again. If the accused was at once able to spit, it boded well for him. The true test came later in the day, when his tongue was carefully examined. If there was swelling or undue burning or swollen glands in the neck, he was declared guilty and paid as his accusers saw fit, often with his life. If he was free of these signs, he was slapped on the back by the soldiers and smiled upon by his king.

At Ubar's gate the king resolved major and minor disputes, gave his blessing to caravans coming and going, and sometimes just passed the time of day. His barber was also his fool. He would dance madly and perform bodily contortions. Or, if the heat was great, he would quietly and slyly compose rhymed jokes at the expense of the Ubarites, the king's retinue, even Khuljan.

A few times a year, lookouts would spot envoys, who had been dispatched to distant nations, now returning. Khuljan received few if any foreign visitors out here; the location of the oasis was best known only to the People of 'Ad. The king preferred to be buffered by middlemen like the Gerrhans to the north, who might take an outrageous cut of the incense trade but would stand in the way of invading armies. Over the years Tiglath-pileser, Alexander the Great, and the Emperor Augustus would have designs on Arabia; for all their might, none would penetrate the incense lands.

This month an envoy might well have returned from Persia, then under the rule of Artaxerxes III. A tribute, this year, had been demanded: a thousand talents' worth of frankincense. Should Khuljan pay it? Dare he defy the Persians?

The influences of Greece and Persia were subtly dividing Arabia, west versus east. Ubar was perilously close to the dividing line. Khuljan and his descendants could well have sided with the Greeks (and later the Romans). Instead, they cast their lot with the Persians, then with their successors, the Parthians. There is a pre-Islamic poem that reflects Ubar's alignment with the east. It describes a journey an envoy would have made, a journey home . . .

> To thee from Babylon we made our way
>> Through the desert wilds o'er the beaten track;
> Oft have our camels from fatigue collapsed
>> And almost failed the distant goal to reach;
> But again they would start with heavy pace
>> To tread the barren route to journey's end . . .
> For Iram of the towers [Ubar] we regard
>> Our sole aim and final destination.[4]

That afternoon, riding about Ubar on his fine horse, Khuljan would have splashed across irrigation ditches and passed through fields of sorghum, millet, wheat, barley, and even indigo and cotton. He would have been pleased by the number of caravans camped in the sprawling oasis. To meet increased demand, there were now two frankincense harvests a year. In the fall and winter months, small, unprotected caravans continuously shuttled the frankincense from the mountains out to Ubar, where it was transferred to larger, armed caravans that departed every few weeks. Two different houses of camels were required for this: animals with smoothly polished, small hoofs bore the incense across the flinty plain leading to Ubar, then animals with large, floppy, soft-soled feet took it across the sands of the Rub' al-Khali. Quite naturally, Ubar's corrals were the logical place to breed and sell dune-adapted camels.

Returning to the fortress, Khuljan would have made his way through the market that every day sprang to life as the sun fell toward the west. Outside the gate, livestock were offered for sale. Camels, the major commodity, were bid for by the casting of stones. Their owners feigned insult at the paltriness of the bids and rhythmically shouted, "The door for more is open! The door for more is open!" Close by, a procession of a dozen goats circled a solitary palm, to be poked and squeezed by potential buyers. A slow-witted man joined the circling goats. At the cost of a dozen stings, he had snatched a honeycomb from a hive in the oasis and was offering it for sale. One by one, each goat buyer broke off a sizable sample, popped it into his mouth, licked his lips, then scowled and shook his head. Not good enough. Before he knew it, the slow-witted man had only his sticky fingers to remind him of his honeycomb.

As Khuljan approached Ubar's gate, the soldiers on duty pushed a playing board out of sight. As he ducked to ride through, Khuljan smiled as the round stones with which the game was played rolled beneath the hoofs of his horse. The fortress's interior courtyard was ringed with stalls. There were merchants of cloth and pottery, merchants hawking olive oil, dried fish, palm beer, and date wine. Sunk to his waist in an earthen pit — for protection from the heat of his furnace — a blacksmith forged arrowheads of molten iron.

A youth screamed. By the northeast tower a curer was at work. Djinns, like men, were drawn to Ubar; the place was infested with them. They weakened the bones but could be driven away by branding. They soured the blood, requiring that it be drawn with heated cups fashioned from the tips of ibex horns. The king paused and watched as a blindfolded youth bobbed and weaved and pleaded for relief from the djinn tormenting him, stealing his vision.

The curer asked, "Are you djinn?"

The djinn — capable of speaking through the mouth of the possessed — didn't answer. Khuljan interjected, "What would you expect? Of course there's a djinn."

The curer said, "Yes, yes, O Lord," and addressed the youth, "You are surely powerful, djinn. What do you want? Tell us. Tell us. Is it gold you want?"

Speaking through the youth, the djinn answered, "A ring."

The curer turned to a knot of the possessed's companions. A ring was reluctantly offered. The curer dropped it into the coals of a frankincense burner, then snatched it up and slid it onto the youth's finger.

Curer: "Djinn, will you remove the evil from the eye?"

Djinn: "Yes."

Curer: "Djinn, swear that you will remove it."

Djinn (its hold lost, its voice choking): "Eh, eh."

Curer: "Be gone!"

Djinn: no answer.

With a sweep of his dirty, blood-stained robe, the curer turned to the assembled and proclaimed, "It has fled. The djinn has fled." The afflicted pulled off his blindfold and began to wobble away, only to be followed, tapped on the shoulder, and reminded, "Gold binds fast the djinn."

Riding on a few paces and dismounting, Khuljan entered Ubar's temple compound. It was as much a house of commerce as a house of the gods. Storerooms and corridors were stacked with sacks of frankincense. Where safer to store it? The temple's garrulous kahin pointed out to the king the measures that belonged to various merchants and those belonging to the temple. Khuljan had the previous year upped the temple's share of the trade from a tenth to a quarter of a caravan's load. The merchants had grumbled and whined, but, as they themselves often said, "The dogs may bark, but the caravan moves on."

We may never know exactly what went on in any temple of ancient Arabia, let alone that of the Ubarites. The identity, nature, and ranking of gods is conflicting and uncertain. It's a mystery which were male and which were female. Temples may have been staffed by regimented orders of priests and priestesses, or they may have been the haunt of soothsayers, even witches.

In Ubar's temple, Khuljan proceeded to a large plastered basin filled with water fresh from the Shisur spring.[5] With a ritual ablution, he purified himself, then mounted the stair to the airless dark sanctuary, the holy of holies, where the gods of his people dwelt in squat stone blocks. These may have been roughly squared off and given suggestions of eyes and mouths, or they may have been uncut. The names of the principal deities of the 'Ad have been mythically reported to be the trio of Sada, Hird, and Haba or the quartet of Sada, Salimah, Raziqah, and Hafizun. Whatever their names, Khuljan would have circled them, chanting an invocation, obsequiously addressing them as masters of Ubar, masters of lands remote and near.

Khuljan was wary of his gods. They, like djinns, could inflict mischief and misery if they were angry, so they had to be kept happy. Sometimes public ceremonies were called for, accompanied by the blood sacrifice of goats and sheep. Today it was sufficient to anoint the stones with oil and offer a burner of frankincense.

As gods brought grief, they also brought benefits. Along with their proper names, they were known as "the rain bringer," "the food-giver," "the savioress," and "the healer." Properly flattered, they would grant benefits in exchange for ritual attention. This year they were in Khuljan's debt, for had he not renovated and enlarged their temple?

This day Khuljan needed a single answer, from the savioress. What should he do about the tribute demanded by the Persians? Was it worth it? Could the 'Ad stand up for themselves? He called for the arrows. The kahin came running with a goatskin bag containing three arrows, each of which had a name: "the enjoiner," "the forbidder," and "the vigilant." The arrows had no heads or feathers, but on one was written "My Lord has commanded me." Another was inscribed "My Lord has forbidden me." The third said nothing. At the king's order the kahin shuffled the arrows and mumbled, averting his eyes from Khuljan, "May you be happy with prosperity and esteem and blessings and — "[6]

Khuljan cut him off as he reached into the bag and withdrew a single arrow. He turned the shaft in his fingers. It was blank. He dropped it back into the bag. The kahin again shuffled the arrows. Khuljan drew again. It was the same arrow, blank. The king's jaw tightened; his eyes narrowed.

The kahin trembled as he shuffled the arrows a third time. He well remembered the time that Khuljan had asked the gods whether to avenge a cousin killed in a dispute over a camel. The king had drawn "the forbidder." Flying into a rage, he had flung the arrow at a sacred stone block and shouted, "You would avenge *your* cousin! Bite your cousin's *zibb!*" Later, though, Khuljan came to his senses, begged forgiveness, and took the unusual step of sacrificing a prize camel in honor of the god that dwelt in the offended rock.

The king withdrew the arrow on which was written "My Lord has commanded me." The soothsayer let out a sigh of relief and said, "The gods know best." Khuljan said nothing and left the temple. He chose to walk rather than ride to the knoll beyond the fortress, where his royal tent was pitched.

Once, centuries before, the religion of the 'Ad may have been more meaningful: it may have had an aura of wandering shepherds reaching for the stars. Once, a temple and its rites may have symbolized the world and its destiny, offering a glimpse of eternity. But no longer.[7] Khuljan and his people were haunted by djinns and consumed by superstition. The gods in their dark chamber were irrational, crass, greedy. Truth be told, Khuljan cared more for his horse.

As it does in the desert, darkness came quickly to Ubar. One by one, oil lamps flickered to life in the king's tent, lit by his fool. The envoy recently returned from Persia awaited Khuljan, nodding gravely as the king announced that the requested tribute would be sent. The gods had ordained it. The envoy thought this prudent and wise, even if the Persian demand was usurious. He had seen for himself the might of Artaxerxes and the splendor of his new palace being built at Susa. It had an oven that could bake an entire ox or camel, so it could be served up whole at dinner. The envoy spoke of what it took and meant to be a Persian king: "an excess of greed, corrupt force, bold daring, momentary success."

Khuljan and his envoy went on to discuss the increasingly complex alliances and enmities of the People of 'Ad. This was not the first mission for the envoy. With his camel stick, he drew a map in the sand and pointed out the territories of rival and friendly kingdoms (see the map on page 232).

In the half of Arabia beholden to the Persians, the envoy noted that the Gerrhans were pirates by sea and brigands by land. Yet the 'Ad were on good terms with them; they were active trading partners. The Rhambanians were a no-account tribe with a puffed-up king. And the people of the Persian Gulf Island of the Two Springs were too distant to matter.

In the half of Arabia under the sway of Greece and Rome, the kingdoms of Ma'in, Saba, and Qataban were too far away to present problems, at least for the time being. It was the increasingly powerful kingdom of the Hadramaut that was troublesome. It was uncomfortably close to the land of 'Ad, and the envoy did not have to remind Khuljan of the adage "If on the trail you meet up with a Hadrami and a deadly snake, kill the Hadrami."

What a puzzle of kingdoms and peoples for such a remote land.

"Enough!" said Khuljan, dismissing the envoy.

The king clapped his hands and called for the wives and children who had accompanied him to Ubar. They were richly arrayed. His wives were unveiled and much freer than in Arabia of later days. Still, they had been painfully marked on their betrothal to Khuljan. He had ordered his fool to make a wide part in their hair by using a razor to remove a strip of skin from their foreheads to the back of their necks. It was a sometimes fatal operation.

Khuljan clapped his hands a second time, and his fool brought forward a gourd of water. The king dipped his right hand in it as if to wash, but did not. Like most of his countrymen, he believed bathing damaged the body. (Although he washed on entering the temple, it was for appearances only.) Thank the gods for frankincense. With a burner, the fool perfumed the king's garments and beard.

The king clapped his hands a third time, and servants brought forth bowls of squash, roasted beans, and meat both raw and roasted. There were flat breads and honeycakes, richly flavored by the nectar of the flowering *elb* tree. There was wine, too, pressed from the grapes that grew high in the Dhofar Mountains. Though the royal company ate well, they ate with haste, a custom born of the ancient reality that mealtime was the best time for lurking enemies to stage a surprise attack.

As the remaining food was cleared away, to be shared by the servants and the king's animals, the fool rubbed the soles of Khuljan's feet and his calves with butter. The king relaxed and whiled away the desert night. From where his tent was pitched, he could take pleasure in looking across at Ubar's fortress, bathed in the light of the moon and set in a diadem of twinkling campfires. Some nights he would send the fool off to recruit camel drivers who could entertain him with their riding chants and songs of memory and love. Other nights the fool would entertain the king's family with jokes and riddles.

"Which is there more of, land or sea?" asked the fool.

"The sea," ventured one of the king's children, "for it goes on forever."

"No," answered the fool, "it is the land, for the sea itself is set upon the land. And what is the sweetest thing in creation?"

"A horse or a camel?" replied the king, only half joking.

"A king's favorite wife," ventured the king's favorite wife.

"Close," said the fool. "The sweetest thing is love from the heart. On this earth it is all we can expect."

On rare and special nights, the king was favored with the presence of a poet. The crafting of verse was considered a great skill, a way to preserve tales of a tribe's history and glory, to immortalize its bold warriors and their dark-eyed women. Poets admitted to being possessed by *shaytan* djinns; how else could they produce anything so complex in rhythm and rhyme, so entrancing?

Enthralled by a woman of the oasis, a poet versified:

> Were it not for her whose wily charms and love
> My heart have captured and my soul possessed,
> Never would I at Iram have pitched my tent . . .[8]

Another poet evoked the melancholy destiny of Ubar and all Arabia: riches may come to you; death will surely come to you. A poem that cites "a man of the race of 'Ad and Iram" might well have portrayed King Khuljan and his court:

> Roast flesh, the glow of fiery wine,
> > to speed on camel fleet and sure.
> White women statue-like that trail
> > rich robes of price with golden hem,
> Wealth, easy lot, no dread of ill,
> > to hear the lute's complaining string.
> These are Life's joys. But man is set
> > the prey of Time, and Time is change.
> Life straight or large, great store or naught,
> > all's one to Time, all men to Death.[9]

The king's fool ventured the riddle: "Who shall conquer all human races?"

"We all know," Khuljan answered. "It is death. Violent and cruel toward all."

When he had had his fill of poetry and wine, Khuljan selected his beloved for the night and prepared to retire. His fool shooed away the other wives and children and extinguished the lamps of the royal tent. All but one. By its light, Khuljan regarded himself in the sheet of polished bronze that served as his glass. Eyes lined with ashen frankincense, how regal was his gaze. He took two wads of cotton with tassels dangling from them, and with his little finger pushed them up his nose, protection against the djinns that rode upon the night air.

The next week or the next month, Khuljan, mighty king of 'Ad, rode away to the mountains and the coast, to Eriyot, his royal city.[10] Over the years he and his heirs enjoyed riches and (as far as is known) remarkable tribal stability. Khuljan and his people, in fact, stood at the threshold of classical achievement, even greatness.

It was a threshold they never crossed.

The 'Ad could have established a formal state, yet instead they remained forever a tribe. They could have created mosaics and heroic statues, yet their vision reached no farther than the rock art on the walls of their caves. The 'Ad could have developed a world view, even a transcendent theology, but instead they worried about lurking djinns and the evils of the night air.

22

City of Good and Evil

THE RISE AND FALL of Ubar spawned a myth of good versus evil. To give it dramatic impact and immediacy, many storytellers have had Ubar destroyed in the very reign of the king who ordered the city's construction. Ubar is barely up before it comes tumbling down. God hardly hesitates before wiping the wicked city from the face of the earth. How better to reward a king who proclaimed, "And people feared my mischief every one."

In reality, following major construction around 350 B.C., Ubar thrived for at least six centuries before its destruction and abandonment. A secret city of frankincense, well fortified, splendid in its isolation. In that era the People of 'Ad enjoyed an advantageous position in Arabia, even as an increasing number of tribes jostled for power. Classical writers called the collective lot of these tribes "Scenitae." Pliny the Elder tells us: "A singular thing too, one half of these almost innumerable tribes live by the pursuits of commerce, the other half by rapine: take them all in all, *they are the richest nations in the world*, seeing that such vast wealth flows in upon them from both the Roman and the Parthian empires; for they sell the produce of the sea or of their forests, while they purchase nothing whatever in return."[1]

To protect their share of Arabia's wealth, the 'Ad aligned themselves with the Parthians, who likely demanded considerable tribute. Yet the Parthians were a long way away when, beginning in the 200s B.C., the People of 'Ad faced increased threats to their control, at its source, of the frankincense trade.

First a migrating tribe, the Omanis, approached from the west and may have threatened Ubar before continuing on their way. Then there was trouble on the coast. Shortly after the time of Christ, the neighboring kingdom of the Hadramaut established a fortified outpost overlooking the best natural port in the land of the 'Ad. They called it Sumhuram, a word likely meaning "the Great Scheme." That it was, for Sumhuram gave the Hadramis control of the sea trade in frankincense. Further, with military efficiency, the Hadramis built facilities for incense collection and storage inland at Hanun and Andhur.

This incursion was not necessarily hostile. The Greeks and Romans now fully understood the seasonal workings of the trade winds and were freely plying the Indian Ocean. The 'Ad may have decided: better an alliance with the Hadrami king, 'Il'ad, than potential conquest by the Romans. It wouldn't be the first instance of a love-hate relationship as old as the Middle East: "Brother against brother, brothers against cousins, brothers and cousins against the world."

Despite what was happening on the coast, evidence suggests that Ubar continued to prosper.[2] What ultimately dimmed its star, and all the stars of Arabia, was a development no one had anticipated: the advent of Christianity. The new religion, as it spread throughout the Middle East, preached that the dead be given a simple burial rather than being cremated, a rite that traditionally called for the burning of enormous quantities of frankincense. For Christians, salvation was gained by belief and good works, not by offerings to the gods. When in 313 A.D. Constantine the Great proclaimed Christianity the favored religion of the Roman Empire, the demand for incense fell off drastically. One by one, the kingdoms of southern Arabia, described by Pliny as "the richest nations in the world," collapsed and were forgotten.

For four years, as myth has it, Ubar was cursed with a drought that withered its crops and killed its animals. If not actual, the drought was metaphorical; the glory days of the incense trade were over. Even so, the king of the 'Ad — now the legendary King Shaddad — was undiminished in his vanity, his arrogance. Shaddad — a name meaning "the strong" — believed himself to be a god, powerful and mighty. The Ubarites agreed. In chorus they proclaimed, "Who is mightier than we?"

To this, one man dissented. He was a handsome merchant, said to be dark-skinned, with flowing hair. He warned of the fate in store for the 'Ad if they persisted in their wicked ways. The man's name was Hud, and he may well have been a Jew, for his name meant "He of the Jews."

It wouldn't have been at all unusual for a wandering Jew to visit Ubar, or even for a faction of the People of 'Ad to have subscribed to Jewish beliefs. Historically, there were several opportunities for Judaism to have penetrated Arabia. As early as the time of Solomon (950 B.C.), Jewish envoys and traders may have traveled the Incense Road. And in one tradition, following their exile to Babylon (587 and 538 B.C.), a contingent of Jews migrated to Dhofar (and Ubar?) and thence to southwestern Yemen, where they quietly survive to this day in the valley of the Wadi Habban. Later, it is certain that in the diaspora precipitated by the Roman conquest of Jerusalem (70 A.D.), numbers of Jews fled to Arabia. Over the years they flourished to the extent that in 520 — the time of the legendary but perhaps real Hud — a Jewish king sat on the throne of a powerful western Arabian kingdom.[3]

At Ubar the stage was set for a morality play, perhaps real, definitely metaphorical. The saintly Hud versus the degenerate Shaddad. Transcendence versus materialism. God versus gods. According to the Koran and subsequent Islamic accounts, Hud was appalled by Shaddad's idolatry; this accords with Islam's tenet that the greatest of all sins is *shirk*, the indiscriminate worship of both lesser beings and material goods. It is uncertain, though, how strongly a real (or even metaphorical) Hud would have felt about this, even if he was Jewish. The Old Testament, though a wellspring of monotheism, directs, "Thou shalt have no other gods *before* me," not "Thou shall have no other gods *but* me." In Hud's era, Judaism in Arabia wasn't all that monotheistic; it appears to have been entranced with the worship of a hierarchy of angels, with the archangel Metatron rivaling the majesty of God.

Evidence of Hud's tolerance of other gods may be found in the story of the delegation of 'Adites that set out for Mecca at his urging to seek relief from Ubar's four years of drought. Mecca then was hardly a center of monotheism; it was, in fact, a swap meet for gods. A pilgrimage often entailed carting a tribal god-block to Mecca and taking another one home in return. The city's holy precinct was choked with 360 tribal idols, complemented by a painting of Jesus and the Virgin Mary. To the Arabians, various deities were sources of power and influence, and it seemed perverse to turn one's back on a potential source of help by opting for only one God.

The prophet Hud may have espoused the worship of El or Allah, a single, transcendent God, and he may have decried the betyls of the People of 'Ad, but there had to be more than that to his quarrel with Shaddad. Consider an obscure but telling fragment of the Ubar legend ascribed to Kaab al-Ahbar. It tells of the palace "which Shaddad ibn 'Ad built and plastered against the wind. . . . When he sat atop his palace with his wives, he would order everyone who passed by, be he who he may, to be killed. God destroyed him."[4] This chilling image raises the question: how wicked were the People of 'Ad, if in fact they were wicked?

Certainly the eye — and the agenda — of the beholder needs to be considered when it comes to wickedness in the biblical era. Nations and tribes (and their chroniclers) have long looked at one another and said, "We can't conquer them, we can't control them, therefore they're ignorant, barbarous, wicked." For all we know, the populace of Sodom and Gomorrah (to say nothing of the entire world before the Flood) may have not been that bad a lot, just a little rough around the edges.

Nonetheless, there was a dark, dystopian side to life in pre-Islamic Arabia. In the works of classical authors and in the inscriptions left behind by southern Arabians, there is a dispiriting sense that life was coarse and brutish, particularly in the Jehiliaya, the approximately four-hundred-year "age of darkness" that preceded the birth of the prophet Muhammad and the rise of Islam. The Arabians were mired in blood feuds and internecine wars. Drunkenness and debauchery were common. The vocabulary of the pre-Islamic Arabians has an astounding number of words descriptive of treachery, cruelty, and malice.[5] HB'Y means "to act corruptly," TBR is "to crush or ruin," RIDH is "to sow death." There appears to be but a single recorded use of the word HMRN, which means "a gracious act."

Every Arabian is by nature "a huckster and merchant," Strabo tells us, and that's the best he has to say. He proceeds to describe a convoluted, dissolute social order:

> Brothers are held in higher honor than children. . . . One woman is also the wife for all, and he who first enters the house before any other has intercourse with her, having first placed his staff before the door, for by custom each man must carry a staff; but she spends the night with the eldest. And therefore all children are brothers. They also have intercourse with their mothers; and the penalty for an adulterer is death; but only the person from another family is an adulterer. A daughter of one of the kings, who was admired for her beauty, had fifteen brothers, who were all in love with her, and therefore visited her unceasingly, one after another. At last, being tired out by their visits, she used the following device: she had staves made like theirs, and when one of them left her, she always put a staff like his in front of the door, and a little later another, and then another — it being her aim that the one who was likely to visit

her next might not have a staff similar to the one in front of the door.[6]

This polyandry arose because of the prevalence of female infanticide. The prophet Muhammad felt the practice was poison to the cup of Arabia. As he sought to reform his world, eliminating it was his first and major social concern. Muhammad's assertions in the Koran are reinforced by a grim account offered by Abu al-Kasim al-Zamakhshari, an early commentator on the Koran:

> When an Arab had a daughter born, if he intended to bring her up, he sent her, clothed in a garment of wool or hair, to keep camels or sheep in the desert; but if he designed to put her to death, he let her live till she became six years old and then said to her mother, "Perfume her, and adorn her, that I may carry her to her mothers"; which being done, the father led her to a well or pit dug for that purpose, and having bid her to look down into it, pushed her in headlong, as he stood behind her, and then filling up the pit, leveled it with the rest of the ground. Others say that when a woman was ready to fall in labor, they dug a pit, on the brink whereof she was to be delivered; and if the child happened to be a daughter, they threw it into the pit; but if a son, they saved it alive.[7]

We can understand why the historian al-Tabari wrote of the "inhuman brutality" of the People of 'Ad, which they "indulged without remorse, and with unmitigated ferocity." So it may have been that, beholding the dark practices of pre-Islamic Arabia, Muhammad preached that Allah told the People of 'Ad: "An ignominious punishment shall be yours this day, because you behaved with pride and injustice of the earth and committed evil."

23

Sons and Thrones Are Destroyed

SOMETIME BETWEEN 300 and 500 A.D., Ubar was suddenly and violently destroyed — both in myth and reality. Over millennia, Ubar's great well had watered countless caravans and had been drawn upon to irrigate a sizable oasis. Handspan by handspan, its waters had receded, and the limestone shelf on which the fortress rested became less and less stable, for it was the water underneath Ubar that quite literally held the place up. If, as in legend, there was a severe drought — and ever more reliance on a single, dwindling spring — the situation would have become critical.

By all accounts, the end came at night. It was likely initiated by a minor tremor, an echo of a faraway earthquake. Yet the seismic shock that hit Ubar was enough to crack and split the limestone underlying the main gate. Almost simultaneously, a huge mass of rock beneath the Citadel gave way, and with a thunderous crash ("the divine shout" of the Ubar legend?) the eastern half of the fifteen-hundred-year-old structure sheared off and plunged into the void below. Anyone inside would have been instantly killed by the crush of tons of masonry and fractured bedrock.

In a few seconds it was over, and a terrible stillness was upon Ubar. A haze of dust rose from the yawning, hellish sinkhole. The colors of that night were the crimson of sudden death, the blackness of the sky, and the pale yellow of the moon. In the broken city, a few shattered oil lamps flickered and died out.

As in its myth, the city had sunk into the sands.

"The next morning," the story has it, "all was ruin." Even so, there would have been survivors, as relatively few people slept inside the city's walls, still preferring the tents of their nomadic ancestors. Terror-stricken, they probably gathered up any treasure that was kept at Ubar and fled across the desert.

At the outset of our search for Ubar, we scarcely imagined that we would find a reality that with a fair degree of accuracy validated the city's myth, but following Juris Zarins's four years of painstaking excavation, it seemed we had. Whether by divine vengeance or the random happenstance of nature, Ubar came to an awful end. For at least the next four centuries, the site's archaeological record — its stratigraphy — tells us that Ubar was a ghost fortress, abandoned.

Yet, Ubar lived on, as we've seen — in memory, imagination, and legend. Following the city's demise, a likely scenario is that the Mahra — a tribe that had its origins in the People of 'Ad — carried the tale of Ubar's fall to the kingdom of the Hadramaut. Then, traveling to Mecca around 610, Bani Zahl ibn Shaitan, a Hadrami merchant, told the prophet Muhammad of the fate that had befallen the wicked 'Adites. Muhammad saw in the story a mirror of the sins of his Meccan opponents and the punishment Allah might have in store for *them* if they continued to ignore and deride him, as the people of 'Ad had laughed at the warnings of Hud.

Once cited in the Koran, the Ubar story was elaborated on by generations of Arab storytellers, threadbare rawis as well as caparisoned court historians. And, possibly even before the revelations of the Koran, the story became part and parcel of Jewish folklore.[1] A Jewish tale has none other than King Solomon visiting ruined Ubar (disregarding the fact that the city was destroyed at least twelve hundred years after his time). It relates how he had a prized piece of tapestry, sixty miles square, on which he flew through the air so swiftly he could eat breakfast in Damascus and have supper in Medina. On one such outing he came to earth in a mysterious desert valley, his attention caught by a great, golden palace. With the exception of a pair of elderly eagles, it was abandoned. The oldest of the birds (aged 1,300 years) recalled that the palace could be entered by an iron door long buried in the sand. In clearing it, Solomon discovered the inscription: "We, the dwellers in this palace, for many years lived in comfort and luxury; then, forced by hunger we ground pearls into flour instead of wheat — but to no avail, and so, when we were about to die, we bequeathed this palace to the eagles."

Passing through the iron door, Solomon wandered through apartments bedecked with pearls and precious stones and confronted a legion of statues that came alive "with great noise and tumult . . . causing earthquake and thunder." He threw them over, and from the throat of one drew a silver plate, inscribed: "I, Shaddad ben 'Ad, ruled over a thousand thousand provinces, rode on a thousand horses, had a thousand thousand kings under me, and slew a thousand thousand heros, and when the Angel of Death approached me, I was powerless."[2]

A further inscription offered a good moral for a bad place: "Whoever doth read this writing, let him give up troubling greatly about this world, for the destiny and end of all men is to die, and nothing remains of a man but his good name."[3]

The real Ubar was not left forever to the eagles. There was still water here, and after an initial period of abandonment the place has been occupied off and on to the present day. Sometime around 900 A.D., Mahra tribesmen rode to the site. With the fortress's old gate collapsed into the sinkhole, they breached the eastern wall so that they could water their horses. Their Arab horses were of sufficient quality to warrant, every year, running a herd across the Rub' al-Khali — via the old Ubar road — to be offered for sale in India. To provide a station for this trade, the Mahra rebuilt walls and parts of the Citadel, but with mud bricks and rubble rather than masonry.

Soon thereafter, someone brought a sandstone chess set to Shisur, as the site was now called, and in a tower of the ruined fortress matched wits in a game that, like the rise and fall of this lonely outpost, ended with the word *shahmaut* — "To the king *(shah)*, death *(maut)*." Or, as we have anglicized it, "Checkmate." The chess pieces were scattered about and forgotten. This may have happened when the fortress was attacked and burned around 940, probably by the Hadramis, who had for a very long time sought to control the 'Ad and, thereafter, the Mahra. In the Citadel, the defenders had stashed but did not use hundreds of iron-tipped arrows (discovered by Juri in 1993). What would the Hadramis have gained? Little but the settling of an ancient score.

In medieval times, Shisur had to have been a melancholy place. In 1221, Ibn Mujawir, a merchant of Baghdad, recorded the final abandonment of the old trade route — the road to Ubar — across the Rub' al-Khali. Travelers Marco Polo and Ibn Battuta wrote of the Dhofar region but said nothing of a city out in the desert (although Ibn Battuta, somewhat sarcastically, mentioned the remnants of the realm of the 'Ad).

The final incident of note in Ubar's history came in the early 1500s when the Hadrami warlord Badr ibn Tuwariq appears to have desultorily rebuilt the old Citadel — and was subsequently credited by the region's bedouin for constructing the entire fortress. The site's true identity — and any hint of all that had happened in and around its walls — was thereafter obscured. Sometime later the bedouin began to believe that Ubar lay hidden in the dunes of the Rub' al-Khali, for that's where they found Neolithic artifacts and that's where the old road ran.

Just when all remembrance of Ubar was fading from bedouin memory (displaced by fascination with a world of Toyotas and Walkmans), an odd chain of events brought an odd collection of adventurers to Shisur. They dug and unearthed an ancient fortress rising above what once was a tent city.

Strata and shards and carbon-14 dates have subsequently given a new reality to the preaching of the prophet Muhammad, to the storytelling of streetcorner rawis, even to the doggerel of contemporary bedouin. A remote desert ruin might have forever remained just that, but for their words . . .

As old as 'Ad . . .

Roast flesh, the glow of fiery wine,
to speed on camel fleet and sure . . .

And ninety concubines, of comely breast
And rounded hips, amused me in its halls . . .

O delegation of drunks, remember your tribe . . .

Wealth, easy lot . . .

An ignominious punishment shall be yours this day, because you behaved with pride and injustice of the earth and committed evil . . .

Sons and thrones are destroyed! . . .

Now all is gone, all this with that . . .

Checkmate . . .

It was a great city, our fathers have told us, that existed of old; a city rich in treasure . . .

At the end of life there is nothing but the whisper of the desert wind; the tinkling of the camel's bell . . .[4]

Epilogue: Hud's Tomb

IN THE SPRING OF 1995, Juris Zarins and his crew wrapped up their archaeological program in Dhofar — and in neighboring Yemen, Kay and I, accompanied by our photographer friends Julie Masterson and David Meltzer, journeyed to where the myth of Ubar came to rest: the tomb of Hud.

Along the way we thought and talked about something that, we saw in retrospect, had underscored our quest for Ubar and the People of 'Ad: *the relationship between myth and landscape.* This relationship has been notably explored by the Australian architect and anthropologist Amos Rapoport, who listened to the stories of Australian bushmen and mapped their world as a mythological landscape. Rapoport perceived that a tribe's cherished myths — of its origin, its meaning, its purpose in the world — were "unobservable realities" that sought expression in "observable reality." Land and landmarks made myth real and validated a tribe and its heritage. To a semi-nomadic, materially rootless people, stories of their ancestors can mean as much as food and water.

Mythological landscapes can be found the world over. There are the kachina-populated mesas and valleys of the Hopi, the Buddhist caves of central China, the landscape of grief and miracles in the Holy Land. As we roamed Oman and then Yemen, it became apparent that southern Arabia had three distinct tiers of mythological landscape. There were the sites of fondly recalled bedouin raids and

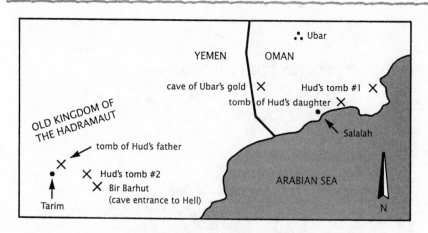

Ubar's mythological landscape

battles, proof of their daring and prowess. Next there were the many and dreaded haunts of djinns. And, lastly, there were sites associated with patriarchs and prophets holy to Islam.[1] This last landscape harked back to the time of Ubar and its principal players, especially the prophet Hud.

Of the two tombs of Hud appearing in this landscape, the oldest and perhaps original one is in an isolated corner of the Dhofar Mountains. It is marked on al-Idrisi's map of Arabia from 1154.[2] We were told by the Omanis that it was a forlorn site and that nobody goes there now, which may or may not be so. In the 1930s, Bertram Thomas wrote of the bedouin perception of the site's awesome power: an oath sworn upon Hud's grave held more weight than one sworn on the Koran or in the name of the prophet Muhammad or of Allah himself. If a guilty party was dragged to the shrine, Thomas recounted, he would usually confess rather than profess his innocence and risk Hud's awful avenging power. Had not Hud brought down the wrath of God on the People of 'Ad?

Regrettably and inexplicably, Omani government restrictions pre-

vented our visiting this site, so David, Julie, Kay, and I found ourselves on our way to a perhaps less authentic, but considerably better
known, Hud's tomb, at the far end of the valley of the Hadramaut in
Yemen.[3] To get there from the capital, Sana'a, we drove through
mist-shrouded highlands, across a land of soaring red rock mesas,
then past a thousand salmon-orange dunes drifting into the sea. On
the fourth day, at the old port of Mukalla, we turned inland and
snaked up onto a drab, featureless tableland. We drove across it for
the better part of a day, until suddenly the ground opened at our feet,
a great rift nearly a thousand feet deep and one to two miles wide.
This was the valley of the Hadramaut, the largest and surely the most
breathtaking wadi in all Arabia.

We descended into a valley of foliage and flowers, surrounded by
the buzzing of bees. (Their honey is unbelievably tasty.) Everywhere
there were neatly furrowed fields tended by black-robed, veiled
women wearing pointed straw hats — witch hats. The Hadramaut's
villages, spaced every few miles, were remarkable: the densely clustered mud-brick dwellings rose four, six, eight stories high. As the last
pale light of the day retreated up the towering cliffs that enclosed
the valley, a thousand and more lamps glowed in the windows of
Shibam, a town where five hundred buildings, soaring 120 feet in the
air, crowded into less than half a square mile. We were in the valley of
the world's first skyscrapers.[4]

In the next few days, we were awed by the valley of the Hadramaut's dramatic setting and fanciful, spectacular architecture. We
were also a little uncomfortable. We had come upon an ancient
way of life, ordered and conservative, with traditions kept very much
to itself. Also, steeped in Ubar lore, I found it hard not to feel that
we were in alien territory, the land of the Hadramis, who had long
threatened and likely pillaged our erstwhile if wicked city. I had to
remind myself that this happened fifteen hundred or more years ago.

We worked our way to the Hadramaut's most distant major settle-

ment: Tarim, city of 365 mosques, once known throughout Islam as
"the city of wisdom and learning." In late medieval times, Tarim had
been celebrated for its great libraries, and we hoped, though we had
been forewarned to the contrary, that one of them might still magically exist and hold a long-lost copy of Ibn al-Kalbi's *History of 'Ad,
the Beginning and the End* or the ten lost volumes of al-Hamdani's
Book of the Crown (documenting the early civilization of Arabia), or
undiscovered early tales of the *Arabian Nights*. But we knew that in
the late 1700s, fanatic Wahhabi tribesmen had overrun Tarim and
ripped to shreds and burned its books. What they missed had been
destroyed by later infestations of white worms. Learning and wisdom
had become ashes and dust.

Even so, there were reverberations of the distant past in the valley's
everyday life. Sacrificial blood consecrated the construction of buildings; white paint splashed around windows warded off djinns; the
social structure of towns, clans, and families was as it had been
hundreds, perhaps thousands, of years ago. And everyone was aware
of the fate of Ubar and the People of 'Ad. The given name Abd
al-Hud — "servant of Hud" — was common in the valley, and leading families claimed Ubar's prophet as their direct ancestor.

Every year Tarim's leading families oversee a three-day pilgrimage
to Hud's tomb, fifty miles to the east. Reportedly, ten to twenty thousand souls travel to the tomb, many of them walking the entire way.
Some oldsters, we were told, hobble a mile or two every year, intent
on eventually accumulating the full fifty. It is a little-known Arabian
event, yet in size and fervor it is second only to the hajj to Mecca. We
would have given anything to witness this but had understood that it
was off-limits to non-Moslems. Indeed, only Moslems who had roots
in the Hadramaut were truly welcome.

We were told, though, that it would be no problem for us to visit
Hud's holy precinct at any other time. And so on an April morning,
awakened by the echoing calls of Tarim's many muezzins, we were

on our way. It was a warm, sunny day, and we were all in good spirits, especially Hussein, our Yemeni driver, who kept time by thumping the stock of his prized Kalashnikov automatic rifle as he sang along with a newly purchased cassette of Hadrami music. East of Tarim the valley of the Hadramaut was quite wide and relatively uncultivated. We passed through several tiny villages but saw only a few distant figures. It was hard to imagine the dirt track clogged with the enormous procession of pilgrims that had passed this way a month before, as they annually did in the second week of the lunar month of Shaban.

By good fortune, there is a detailed account of the pilgrimage to Hud's tomb, written by Arabist-anthropologist Robert Serjeant in the late 1940s. As he described it, the procession out across the desert from Tarim was a high-spirited, often boisterous affair, with the pilgrims frequently breaking into song. There was the song of the cloud that accompanied Hud on his travels, ever shielding him from the sun. There was the song of the trapped gazelle who asked Hud to free her so she might feed her young. There were quaint songs, insulting and bawdy, for the villages marking the way. One ditty deemed the rock-hard ground at As-Sallalah uncomfortable for sleeping, fit only for buggery:

> You camel men, go, abuse each other,
> At As-Sallalah go meet your lover.

Passing through the tiny town of Khon, the pilgrims sang:

> O Khon, no girl in Khon is chaste,
> Where married and unmarried women fornicate.[5]

Through As-Sallalah and on through Khon we drove. We turned off into the Wadi 'Aidid, a tributary of the Wadi Hadramaut. Far ahead, set on a rise at the foot of a towering red rock cliff, we caught sight of a glistening white dome. As we drew closer, we saw that the tomb overlooked a fair-sized town, which proved to be immaculately well tended and clearly prosperous — but totally deserted. No dog barked, no bird sang, no retainer looked after things. It was inhabited but three days a year, during the pilgrimage.

From the white, silent town, a broad staircase led up to an open-air prayer hall built around a giant boulder. A further flight of stairs ascended to the tomb itself, a graceful dome in which was enshrined a whitewashed, squared-off rock, split by a fissure. It was the sarcophagus of the prophet Hud. He was evidently a very tall man, for his sarcophagus extended beyond the confines of the building and a good ninety feet up the hillside behind it. The tomb's walls were speckled with colored dots, wads of paper conveying the prayers and pleas of recent pilgrims.

Though augmented by buses and pickups, this year's procession from Tarim, we had been told, was much as Serjeant had described it in the 1940s. Passing an outcropping called the Rock of the Infidel Woman, pilgrims shouted, "God curse you, infidel woman," and, better yet, peppered it with bursts of rifle fire. Their arrival at the town had prompted chanting, shouting, and more shooting. The atmosphere was festive, harking back to the great fairs of pre-Islamic Arabia, where the people of the desert and the cultivated lands, herdsmen and farmers, gathered to exchange goods, race camels, and honor not only Hud but, as a present-day Hadrami has written, "invoke peace on all the prophets of importance, the four perfect women, the archangels, and the gardener of Paradise."[6]

Close by holy Hud's town, water gushed inexplicably from the barren desert (a phenomenon that may have made this a place of pilgrimage long before its identification with him). The devout made their ablutions here, in the belief that they were bathing in the waters of a river of Paradise. Then they joined a long, slowly moving line that ritually retraced Hud's last days.

Hud, they believed, was pursued into the Wadi 'Aidid by a pair of wild and godless horsemen. He rode up the gully beyond the town (where there is now a wide staircase) and leapt from his faithful she-camel — a beast that God immortalized by turning it into a great boulder. Cornered and hard-pressed, Hud said to an oblong rock before him, "Open by the permission of God!" The rock opened wide. He entered, and the rock closed behind him, though it did not close entirely. A fissure remained through which, it is said, only the virtuous can pass.

Though pilgrims have demonstrated their virtue by squeezing partway into the fissure, what lies beyond it is holy, not to be seen. In the 900s, the Yemeni historian al-Hamdani offered the report of an informant from the Hadramaut: "As I went in, I saw stretched on a bier a man of dark brown complexion with a long face and thick beard. The corpse was dried up and felt hard to the fingers. I saw beside his head the following inscription in Arabic: 'I am Hud who believed in God. I had compassion upon 'Ad and regretted their unbelief. Verily nothing can forestall what God has ordained.'"[7]

In this far corner of far Arabia, we had come to a precinct that pilgrims held "so sacred that a stick or stone removed would come alive, leaping and screaming until it was replaced."[8] We were at the center of a vivid mythological landscape. It encompassed belief (Hud's tomb) and unbelief (the Rock of the Infidel Woman). It encompassed heaven and hell; across the Wadi 'Aidid were pools watered by a river of Paradise, and a three-hour walk away was the cave and well of Bir Barhut, widely believed to be a sulfurous portal to the underworld.[9]

Journeying to Hud's tomb, the pilgrim entered a symbolic world of the past, the psyche, life, and death. The pilgrimage lamented the death of the prophet Hud in a landscape of death; the word "Hadramaut" — appearing in Genesis as Hazarmaveth — has been taken to mean "Valley of Death." Yet this landscape also served as a landscape of life, of fertility. Robert Serjeant tells of the Karat Mawla, a conical (phallic, he says) hill in the Wadi 'Aidid. If a woman does not become pregnant within two or three years of marriage, she may elect to climb to the top of this hill, strip herself naked, and lie on her back as if anticipating intercourse. In some cases, the husband is there and materially enhances the chance that the woman will become pregnant. A pilgrim confided to Serjeant succinctly, "Some people have tried this out and benefited."

If a curious and diligent researcher could have unrestricted access to Hud's contemporary pilgrims and their pilgrimage, there would be enough material for a major study of ancient belief and rites, transmitted from the time of belief in betyls to the time of Islam. We were content just to be here and to sense the abiding power of Hud, prophet of Ubar. As a social reformer, he challenged a people who may have been not only "arrogant and unjust" but who had probably fallen into the barbarism of infanticide and Lord knows what else.

As to Hud's condemnation of the worship of multiple gods, the question can be raised: what, inherently, was wrong with worshipping as many gods as one wished? (In the old days, who didn't?) But consider the nature of Arabia's gods: they were identified with celestial and natural forces (scorching sun, comforting moon, storms that could be either destructive or life-giving). Offerings were made to please them, to gain their favor. Whether the petitioner was avaricious or dissolute didn't matter. The gods had little or no interest in morality. By contrast, a single God, particularly a Jewish-inspired single God, historically called for the judgment of human behavior. What mattered in life came to be moral order and elemental human decency. Faith might be important, but decency was even more important. So it has been said in Arabia, by pious prophets and free-spirited bedouin alike: when the vanity of the world fades and is gone, nothing remains of an individual but his good name.

For Kay and me, this was the day we reached the end of a fifteen-year trail, for it was that long ago that Virginia Blackburn, the crusty bookseller in Los Angeles, had insisted I buy a book I didn't want to buy. A few nights later Kay and I first came upon the story of the ancient lost city of Ubar. Between then and now we had had many doubts whether Ubar, much less Hud, really existed, and in pursuing this quest, had often been but a step ahead of the "We would like to remind you . . . perhaps you overlooked . . ." people at American Express and Visa. But now the search was, as they say in Arabic with a clap of the hands, "Khalas!" Finished. "Khalas!" an Omani good friend told us, has a double meaning. As well as "Finished!" it means "Salvation!"

As we wandered about exploring the environs of Hud's tomb, our driver, Hussein, napped in the shade of Hud's petrified camel, his Kalashnikov his pillow. At our return, he blinked awake and asked, "We go?" We bumped over the desert track back to Tarim, where we spent the night at a derelict palace now being run as something resembling a hotel. It was, lugubriously but appropriately, named the Qasr al-Qubba, the "Castle of the Grave." A sign over the front desk read: "All weapons to be left with the Management." Hussein said if that was the case, he would just as soon sleep on the roof of his Toyota (and did). On the way to our rooms, we passed another sign, lettered in Arabic, then English:

CALMN-
-ESS IS
REQUE-
-STED
FROM
ALL.

The rooms were stifling hot and not for the fastidious. We hesitated to open the windows for fear that creatures of the night might join us. Already a suction-footed lizard was adhered to a pane, watching us. Or perhaps he was more interested in the little snacks our room might offer. A creature of uncertain species strolled across our pillow, which reminded us of the enlightenment offered by a Middle Eastern hotel clerk on a previous occasion. "The reason there are bugs in the bed," he explained, "is that they're too scared to get down on the floor."

As darkness descended over the Hadramaut — the Valley of Death — we settled in for the night at the Castle of the Grave. We didn't sleep badly. Quite well, in fact.

Key Dates in the History of Ubar

1,000,000–100,000 B.C.	*Homo erectus* in the vicinity.
100,000–20,000 B.C.	Migrating from Africa, *Homo sapiens* camps at Shisur spring (the nexus of what would later become Ubar). In this era, Arabia is a vast savanna.
20,000–8000 B.C.	A devastating era of hyperaridity turns Arabia into an uninhabitable wasteland.
8000–2500 B.C.	The rains return, and with them pastoral nomads who construct a large animal trap at Shisur. They harvest frankincense and conduct long-range trade with Mesopotamia.
2500 B.C.–present	The rains retreat, initiating a new period of aridity that continues today.
c. 2000 B.C.	The camel is domesticated, possibly in southern Arabia.

c. 900 B.C.   Ubar's Old Town built.

c. 350 B.C.   Ubar's New Town built. Trade extends to
                         Egypt, Israel, Greece, and Rome. Ubar's
                         days of glory (and perhaps inglory) follow.

c. 300–500 A.D.   Ubar destroyed and abandoned.

900–1500   Ruins of Ubar reoccupied; minimal rebuild-
                         ing. (Evidence of attack and burning,
                         c. 940.)

1930   Explorer Bertram Thomas discovers "the
                         road to Ubar." In Thomas's footsteps, expe-
                         ditions seek the city in 1932, 1945 (two at-
                         tempts), 1953, 1956, and, finally, 1991–92.

A Glossary of People and Places

'AD   The people who in antiquity harvested Arabia's finest frankincense from groves high in the Dhofar Mountains of today's Oman. Ubar was the 'Ad's city in the desert.

AIN HUMRAN A fortress of the 'Ad overlooking the Arabian Sea and controlling the maritime shipment of frankincense. In architecture and purpose, it was Ubar's sister city.

ANDHUR A colonial outpost of the Kingdom of the Hadramaut in the territory of the People of 'Ad. Along with Hanun, it was an inland collection point for frankincense.

AL-AHQAF An arc of dunes on the southern edge of the Rub' al-Khali. In legend, this is where Ubar lay buried.

DHOFAR The southern region of today's Oman, where the Dhofar Mountains rise up from the coast, providing ideal conditions for the growth of Arabia's finest frankincense.

GERRHA A city on the north side of the Rub' al-Khali. The Gerrhans were trading partners of the 'Ad.

HAGIF The 'Ad's major settlement in the Dhofar Mountains and the largest Bronze Age site in Oman.

HADRAMAUT A powerful kingdom immediately to the west of the People of 'Ad. Shortly after the time of Christ, the Hadramis sought a share of the frankincense harvest and colonized 'Adite territory.

HANUN A colonial outpost of the Hadramaut in 'Adite territory. Along with Andhur, an inland collection point for frankincense.

HUD In legend, the prophet who warned the People of 'Ad of the terrible fate that would befall them if they failed to renounce their arrogant and wicked ways.

IRAM A name for Ubar in the Koran, the *Arabian Nights*, and many other accounts.

KHOR SULI An 'Ad port on the Arabian Sea for the shipment of frankincense. In case of attack, its inhabitants could retreat to the nearby fortress of Ain Humran.

KHULJAN In legend, the greatest king of the People of 'Ad.

MAHRA A desert tribe descended from the People of 'Ad that exists to this day.

OMANUM EMPORIUM The apparent designation for Ubar on Claudius Ptolemy's map of Arabia, 150 A.D.

RUB' AL-KHALI (THE EMPTY QUARTER) The great sand desert of central Arabia, the largest sand mass on earth.

SABA (OR SHEBA) A famed Arabian kingdom far to the west of Ubar. Known for its queen, who journeyed to Jerusalem and the court of King Solomon.

SHADDAD In legend, an 'Adite king known for his arrogance and vanity.

SHAHRA Today a small tribe living in the Dhofar Mountains. The Shahra claim direct descent from the People of 'Ad.

S H I S U R The spring at Ubar and today's name for the site.

S U M H U R A M The principal colonial settlement of the Hadramaut in the land of the 'Ad. A port for the shipment of frankincense collected at Andhur and Hanun.

U B A R The legendary "Atlantis of the Sands," the city doomed to destruction because its people "sinned the old sins, and invented new ones." In reality, a staging point for the caravans bearing frankincense north to Mesopotamia, Egypt, Israel, Greece, and Rome. In both myth and reality, Ubar was destroyed in a great cataclysm.

W A B A R A variant spelling of Ubar.

APPENDIX 3

Further Reflections on al-Kisai's "The Prophet Hud"

"The Prophet Hud," as told by Muhammad ibn Abdallah al-Kisai, is the work of a master storyteller. He casts the tale of Iram/Ubar as a three-act structure, with each act broken into short scenes. The structure arose out of the need of a *rawi*, a street storyteller, to be paid for his prose. Every so often he would pause at a point where onlookers were anxious to know "what comes next." The rawi would nod to his *mukawwiz* (collector) that now was the time to rattle his cup and take up a collection.

On one level, "The Prophet Hud" is a morality play with a flair for fantasy. On another, its subtext is steeped in fascinating details of life before Islam, including choice clues as to the character and location of Iram/Ubar. Virtually every aspect of the tale can be traced back in time; its every thread has a source. Little, if anything, is woven of whole cloth.

A number of the story's telling ideas, citations, and characters are discussed in Chapter 7, "The Rawi's Tale." Here are more. The numbers refer to the lines on pages 81–87.

Line 5: "Wahb ibn Munabbih said: The greatest king of 'Ad was Khuljan . . ."
Line 44: "Kaab al-Ahbar said: When Hud was four years old, God spoke to him . . ."

To enhance their credibility, Arab storytellers cited past

chroniclers, sometimes by the score. But "The Prophet Hud" mentions only three, and two of these — Ibn Munab-bih and al-Ahbar — had a common agenda. Both were Jewish converts to Islam, and both were anxious to prove that their new religion shone as the only true faith. Nevertheless, they valued their Jewish heritage. Fond of its figures and folklore, they would have been particularly taken by the idea of a Jewish prophet Hud.

Line 6: "three idols, Sada, Hird, and Haba . . ."
Here we have a glimpse of the pagan Arabia that Hud decried, with its religion centered on a celestial trinity of moon, sun, and morning star.

Lines 11–18: "I saw coming out of my loins a white chain . . ."
Aside from this passage's rambunctiously sexual imagery, a chain in early Islam bespoke power and control. It was often associated with the execution of divine will, whether beneficent (as here, heralding Hud's conception) or vengeful (in hell sinners are not only wrapped in chains but skewered on them, to be "roasted on fire as 'kabob' is grilled on skillets"). Ibn Kathir quoted in Khawaja Muhammad Islam, *The Spectacle of Death* (Des Plaines, Ill.: Kazi Publications, 1987), p. 301.

Line 25: "he was born on Friday."
Propitiously, Hud is born on the day of the week holy to Islam. The stage is set for a contest between adherents of many gods and a believer in one God.

Lines 33–44: "'My child,' she said, 'worship your god, for on the day I conceived you I saw many strange things.'"

Hud's mother recalls six wondrous signs, which offer a sampling of the folklore of ancient Arabia. The black rock miraculously turned white is an inversion of the tradition that the black rock set in the corner of Mecca's holy shrine (the Ka'aba) was once dazzlingly white, darkening only in the shadow of the sins of man. Symbolically, Hud can reverse this and lead his people out of darkness into the light.

There are other images of light and white. The people of the heavens have white faces; rays of light bless the infant Hud. Then, as a final and inspired touch, a pearl, which is normally white, appears as a sign on Hud's arm. But here this talisman is green, the color of Islam.

The giant man who lifts Hud's mother into the sky harks back to the tradition that the People of 'Ad were a race of giants. Here is an excellent example of the creative role of exaggeration in Arabian folklore. Where we might employ relatively unimaginative words such as "impressive" or "beyond belief," Arab storytellers would describe a city as encrusted with rubies and pearls, or add a few zeroes to a tribe's population. An idea that was beyond man's comprehension would be symbolized by a being that was beyond comprehension; in this case a vision of heaven is delivered by "a man whose head was in the sky and whose feet were in the vast expanses of the earth." And this giant is a little guy compared to Arabian visions of angels: Gabriel (who appears later in the tale) is traditionally described as having a thousand eyes, wings that cover the earth, and a face that radiates the light of a thousand suns. The point is that the beings who flank God in his heaven — to say nothing of God himself — are beyond measurement in human terms (and infinitely more imposing than squat stone idols).

Lines 26–27: "his mother saw him and asked, 'My son, whom are you worshipping?' . . ."

Lines 44–77: "When Hud was four years old God spoke to him . . ."

The prophet Hud's relationship with his God — and his people — is at first low-key. As a small child, he mildly rebukes idolatry by pointing out to his mother, "These idols bring neither harm nor profit . . . Neither do they see or hear." They are useless blocks of rock. But then the situation escalates. Fired by a message from God, Hud challenges his people — and his king — to worship but a single God. He wins a few converts but otherwise is rebuked and cursed. He counters with a seventy-year series of warnings and threats. To no avail. Even when the wrath of God comes down upon the People of 'Ad, they turn a deaf ear to Hud, his chosen prophet.

Regrettably, this arc of character and story development probably has little to do with actual events at Ubar. Rather, it is about a different man in a different age. It is an extraordinary replication — rich in emotion and detail — of the early career of the prophet Muhammad. Confronting the malaise and materialism of the people of Mecca, Muhammad became increasingly frustrated and angry at their disbelief, only to have his own tribe, the Quraish, proclaim his Koran a forgery and reject him as an impostor. As he walked the streets of Mecca, he was showered with insults.

It was then that, like Hud, Muhammad took it upon himself to become a "warner." His sermons, in fact, repeatedly brought up the story of Iram/Ubar as a prime example of the fate God had in store for the Meccans. It was only when he was in Mecca that Muhammad delivered verses of the Koran dealing with the People of 'Ad. Agitated, intense, inspired, they have been called the "Terrific Suras."

It is a little confusing, but the dynamic here is three-tiered: our tale's Hud (described in the 1100s) is modeled on the prophet Muhammad (600s), who in turn equated himself with a historic Hud (150–500)!

During his years in Mecca, Muhammad would have had no problem likening himself to a Jewish prototype. It was only after he left Mecca and migrated to Medina that he had a serious falling-out with the Jews. Even then, he honored them as "the People of the Book," that is, the Old Testament.

Lines 98–133: "it was the custom, when a people was afflicted from heaven or from an enemy, to take an offering to the Sanctuary of the Ka'aba . . ."

In the Koran, there is no mention of a delegation from Ubar/Iram making a pilgrimage to Mecca. This has apparently been added to "bring home" the story by having the 'Ad, who gave the prophet Hud grief, travel to the city that gave the prophet Muhammad grief.

Here also is a glimpse of a pre-Islamic Meccan pilgrimage. The seventy chosen men enter the Sanctuary on jeweled she-camels, and there is a rite involving the draping of robes.

Line 104: "and their names were Qayl, Luqman . . ."

The inclusion of the name Luqman connects the Iram/Ubar story to a vast web of interrelated Arabian legends. Luqman, it is written elsewhere, was granted the lifespan of seven generations of captive vultures; he wanders the Middle East for 650 to 3,500 years (depending on what source you read and what the author considered a vulture's lifespan).

Our rawi's tale signals Luqman's very first, understated appearance. Unlike his fellow delegates to Mecca, he has nothing to say or do. In years and legends to come, he makes up for it. He appears in Arabian and African tales as a vagabond, a shepherd, a deformed slave, a tailor, a carpenter. He composes proverbs and fables. He has the intellect of a hundred men and is the tallest of all. He becomes vizier to King David, who considers himself fortunate and proclaims: "Hail to thee, thine is the wisdom, ours the pain!" He becomes a king himself, king of 'Ad the Second, a realm equated with the city-state of Sheba. There he builds the Great Dam of Marib, which makes it onto several lists of "Wonders of the Ancient World." (Its monumental, ruined masonry is still to be seen.)

When the last of the vultures reared by Luqman finally falls off the perch, Luqman stirs him to fly again, but in vain. The bird dies, and Luqman with him. The name of this last vulture is Lubad — "Endurance."

What a life, what a sustained flight of fancy! Like a genie, a good character let loose is hard to put back in the bottle, and there is no telling what he'll become. This is not to say that a Luqman never existed, only that his early incarnations — as a desert vagabond or shepherd — may have been closer to the truth. (By the same token, Luqman's first Arabian haunt — Ubar — could have been a relatively modest settlement.)

Lines 121–129: "he sent them two slave-girls, called the Two Locusts, who were singers in his service . . ."

Pairs of singing girls were a staple of pre-Islamic entertainment. What's interesting here is that the Locusts are decidedly free-spirited (a contrast to the reclusive stereotype of women in Arabia today). With little inhibition, they mock their audience of out-of-towners.

Lines 149–173: "God's angel Gabriel said, 'O cloud of the Barren Wind, be a torment to the people of 'Ad and a mercy to others!'"

With these words, a torrent of imagery is let loose. And here we can imagine a blind medieval rawi, on the steps of a Cairo mosque, building to his tale's apocalyptic climax. It is late in the evening. Merchants have shuttered their stalls, yet people are abroad, seeking the breeze that comes on the wings of night. They're drawn to the rawi, who melodramatically lowers his voice as he relates: *"On the first day, the wind came so cold and gray that it left nothing on the face of the earth unshattered."* The eyes of little boys at his feet widen. The better to hear, the crowd presses in. *"On the second day there was a yellow wind that touched nothing it did not tear up and throw in the air."* The rawi melodramatically pauses and gropes to light an oil lamp; its flicker eerily brings life to his lifeless eyes. *"On the third day a red wind left nothing undestroyed."* He talks faster now, mimicking the cry of the defiant 'Adites: *"We are mightier than you, Lord of Hud!"* The rawi now shouts, louder than anyone could imagine, stunning his audience: *"The wind ripped them apart and went into their clothing, raised them into the air and cast them down on their heads, dead."* The rawi's imagery is increasingly fervid, gruesome — and powerfully poetic. With grim finality the rawi seals his story: *"Sons and thrones are destroyed!"*

Looking on, shaking his head, the rawi's mukawwiz mutters "Iram, khalas" . . . Iram, finished. The crowd sighs in relief and appreciation. The blind rawi faintly smiles as the mukawwiz's cup rings with dinars.

This climactic passage is packed with derivations, allusions, and lore. For example, our writer-rawi was no doubt familiar with the prophet Muhammad's antipathy to arrows, which in pagan Arabia were instruments of both gambling and divination. So when the 'Adites defy God's wind by shooting arrows at it, "the wind snatched their arrows and drove them into their throats." And the bizarre-seeming notion that the wind entered Khuljan's "mouth and came out his posterior" reflects an Arabian belief — still heard today — that the body is hollow. The concept of dying from a faceful of wind goes back to the *Enuma elish*, the Babylonian creation myth in which a wind is driven into the goddess Tiamat's mouth; it gruesomely distends and destroys her.

Considered as a whole rather than as the sum of its parts, this passage is the payoff of a powerful myth. In the vision of mythologist Joseph Campbell, the essential function of myth is to pull individuals into accord with the universe. In warning the 'Adites, that is exactly what the prophet Hud tries to do. They could not care less; they revel in materialism, ignoring God. If anything, they consider themselves *above* any cosmic order. With the end clearly in sight, the 'Adites still scream, "We are mightier than you, Lord of Hud!"

The response, of course, is: God is mightier than you. And He proves it, wiping the 'Adites from the face of the earth.

But not all of them. The tale takes pains to add that Hud and a number of his followers survive, so that (as Joseph Campbell would have it) they may pursue, unhindered and anew, an accord with the universe.

In the world of early and medieval Islam, the story arc of *sin, then warning, then more sin, then punishment* was by no means unique to "The Prophet Hud." When it came to moral weapons, Muhammad enthusiastically chose the fear of God. In the Koran, time and time again, he tells of prophets spurned and cities and civilizations consequently destroyed by an angry God. The pattern goes back to Adam, who in Islam is not only a progenitor but a prophet. Speaking from his own recent and humbling experience, Islam's Adam instructs mankind in the correct way to live. But mankind never quite gets the message — despite the bad end that comes to Sodom and Gomorrah, despite the onset of catastrophes foretold by Noah, Joseph, Hud, Saleh, even Jesus (Isa in Islam).

Lines 178–198: "Kaab al Ahbar said: One day I was in the Prophet's Mosque . . ."

Though the curtain has inexorably rung down on Ubar/Iram, there is more to the story. Adhering to good dramatic form, the climax of "The Prophet Hud" is followed by an anticlimax, an epilogue that eases the reader (or listener) back to the present. Moreover, the reader is assured that indeed there was such a place as Ubar, such a prophet as Hud. The evidence offered is Hud's tomb in Yemen's valley of the Hadramaut.

To this day, Hud's tomb is the most popular pilgrimage site in southern Arabia. Throngs of pilgrims offer incense at "the opening through which a thin man may pass." And they know well the story of the wicked city that denied the message of God's apostle Hud. Ubar may have been wiped from the face of the earth, but it was not — and is not — forgotten.

Notes

Prologue

1. "When I had finished reading the book . . .," Rev. Mr. J. Cooper, trans., *The Oriental Moralist or the Beauties of the Arabian Nights Entertainments* (Dover, N.H.: Printed by Samuel Bragg, Jr. for Wm T. Clap, Boston, 1797), p. i.
2. "That God holds you over the pit of hell . . .," Clarence H. Faust and Thomas H. Johnson, eds., *Jonathan Edwards* (New York: American Book Co., 1935), p. 164.
3. "We set sail with a fair wind . . .," "Exploring the town's fantastical palace . . .," "It was about three years ago . . .," and "Since that time I have whipped them . . .," Cooper, "The Petrified City," *Oriental Moralist*, pp. 163–74.
4. "The Arabs were an ignorant, savage and barbarous people . . .," J. Olney, *A Practical System of Modern Geography* (New York: Robertson, Pratt, 1835), p. 201.

1. Unicorns

1. "one horn in the middle of his forehead . . .," T. H. White, *The Book of Beasts: Being a translation from the Latin Bestiary of the Twelfth Century* (London: Jonathan Cape, 1954), pp. 20–21.

2. *The Sands of Their Desire*

1. "right foul folk and cruel . . .," John Mandeville, *Mandeville's Travels*, vol. 1 (London: Hakluyt Society, 1953), p. 47; "The people generally are addicted . . .," William Lithgow, *The Totall discourse of the Rare Adventures and Painefull Peregrenations of a Long Nineteene Years Traveyles from Scotland to the Most Famous Kingdoms in Europe, Asia and Africa* (1612; reprint, Glasgow: J. MacLenose, 1906), p. 262.

2. "dashed down the mountain . . .," John Lloyd Stephens, *Incidents of Travel in Egypt, Arabia Petraea and the Holy Land*, vol. 2 (New York: Harper & Brothers, 1837), p. 12.

3. *he was a simple "Jewish Jesuit."* What a piece of work was Gifford Palgrave. During his years with the Jesuits, he assumed the names of Michel Sohail, Michael X. Cohen, and Seleem Abou Mahmood el Eys. In Arabia, while still insisting he was a "Jewish Jesuit," he allowed that he had been "invested for the nonce with the character and duties of an Imam, and as such conducted the customary congregational worship." (Palgrave, "The Mahometan Revival," in *Essays on Eastern Questions* [London: Macmillan, 1872], p. 126.) In his later years, he was enamored of Shintoism.

4. "Nothing but an airship can do it," David Garnett, ed., *The Letters of T. E. Lawrence* (London: Jonathan Cape, 1964), p. 663.

5. "'Why aren't you married, O Wazir?' . . .," Bertram Thomas, *Alarms and Excursions in Arabia* (Indianapolis: Bobbs-Merrill, 1931), p. 119.

6. *Harry St. John Philby.* If the name Philby seems familiar, it is probably because of Harry's son, Kim, the well-known KGB mole. With a craftiness that may have run in the family, in the 1950s Kim Philby infiltrated not only Britain's Secret Intelligence Service but, as a liaison officer, the CIA.

7. "the sands of my desire," Bertram Thomas, *Arabia Felix: Across the "Empty Quarter" of Arabia* (New York: Charles Scribner's Sons, 1932), p. 149; "the bride of my constant desire," Harry St. John Philby, *The Empty Quarter* (New York: Henry Holt, 1933), p. xxi.

8. "Tomorrow, the news of my disappearance . . ." This and the following quotes are from Thomas, *Arabia Felix*, pp. 1, 2, 42, 131, 136, 149.

9. *with an accuracy that is amazing.* For decades Bertram Thomas's map would be the only reliable guide to the Dhofar region of Oman. When we began looking for Ubar from space, I superimposed portions of Thomas's map on satellite images and found that he was never more than a kilometer or two off in plotting his route.

10. "Our morning start was sluggish . . .," Thomas, *Arabia Felix*, pp. 160–61.

11. "'Twas I that learn'd him in the archer's art . . .," Zayn Bilkadi, "The Wabar Meteorite," *Aramco World Magazine* 37, no. 6 (Nov.–Dec. 1986), p. 28.

12. "the finest thing in Arabian exploration . . .," T. E. Lawrence, foreword to Thomas, *Arabia Felix*, pp. xix, xvii.

13. "Here then the words of 'Ad . . ." This and the following quotes are from Philby, *Empty Quarter*, pp. 165–66.

14. "mantle of fraud in the east . . .," T. E. Lawrence, *The Seven Pillars of Wisdom* (Garden City, N.Y.: Doubleday, Doran, 1936), p. 503.

15. "I am convinced that the remains . . .," Raymond O'Shea, *The Sand Kings of Oman* (London: Methuen, 1947), pp. 180–81.

16. "A cloud gathers . . ." and "this cruel land can cast a spell . . .," Wilfred Thesiger, *Arabian Sands* (New York: E. P. Dutton, 1959), p. xvii. Thesiger describes (p. 219) a typical Ubar discussion: "According to Sadr . . . the lost city of 'Ad [was] under the

sands of Jaihman. Muhammad was, however, convinced that this city, one of the two mentioned in the Koran as having been destroyed by God for arrogance, was buried in the sands to the north of Habarut. He reminded me of the many clearly defined tracks which converge on these sands, and which the Rashid maintain once led to that city."

17. *Had he sought Ubar . . . ?* Years later, I listened as one of Thesiger's bedouin companions, Sultan Najran, relived the journey. I learned that Thesiger, indeed, had looked for — and found — the road to Ubar, but at the cost of draining his party's waterskins dry. His party was lucky to make it back alive.

18. "Qidan, the lost city . . .," O'Shea, *Sand Kings of Oman*, p. 1.

19. "I realized with a start . . .," James Morris, *Sultan in Oman* (London: Century Travellers, 1986), p. 121. Morris's book is a wry and immensely entertaining portrait of the world of Sa'id ibn Taimur, the sultan who had once retained Bertram Thomas as his wazir.

20. *His mysterious mesa might well have been a desert outpost . . .* The site of "Qidan" could in reality be Muscalet, a settlement that appears on maps from the 1600s into the 1800s. Or it could indeed be ancient, the "Rhabana Regia" of Ptolemy's venerable map of Arabia. In an image taken by the high-resolution Large Format Camera aboard the space shuttle *Challenger*, O'Shea's mesa is where he said it was and has linear features that could well be man-made.

21. *Mahram Bilqis. Mahram* in Arabic means holy platform or sanctuary, and Bilqis is a traditional proper name for the queen of Sheba. In Yemen today, if you shout "Bilqis," half the little girls within earshot come running.

22. "almost trampled over the rest of us . . ." This and the following three quotes are from Wendell Phillips, *Qataban and Sheba* (New York: Harcourt, Brace, 1955), pp. 225, 264, and 307.

23. "When I enquired if he knew . . ." This and the following quotes

are from Wendell Phillips, *Unknown Oman* (London: Longman, 1966), pp. 222–23, 223–24, and 229.

24. "It was California Charlie . . ." Charlie McCollum, the fellow
who spotted the Ubar road, was a Phillips sidekick. I managed to
track him down in California; he confirmed that the road was as
wide as a ten-lane freeway.

25. *He shouted "This is Ubar!"* Phillips's bittersweet jest does not
appear in his *Unknown Oman*. It was related to me by one of his
guides, Muhammad ibn Tuffel.

26. *and prospered in the oil business.* As a gesture of courtesy, the
desert sheiks offered Phillips their homes, their possessions, anything he wished. He responded, no, no, they were too kind. All
he asked was that they sign an option for the oil rights to their
tribal lands. Brokering these options, Phillips became the world's
leading private oil concessionaire — and his desert friends prospered as well.

3. Arabia Felix

1. "He is crazed with the spell . . .," from "Arabia," Walter de la
Mare, *Collected Poems*, vol. 1 (New York: Henry Holt, 1920),
p. 135.

2. *none had really done his homework.* Freya Stark, a hardy solo
traveler of the 1930s, was apparently the only one to comb ancient accounts. But she never took Ubar seriously and never
went looking for the city, even though her friend the sultan of
Qatn "told me in the serenity of faith that everyone in Hadramaut places it between Hadramaut and Oman" (where
Thomas, Thesiger, and Phillips encountered the Ubar road).
Freya Stark, *The Southern Gates of Arabia* (Los Angeles: J. P.
Tarcher, 1983), p. 181.

3. *the land was uncharted.* Reliable maps of Arabia were a long
time coming. As late as World War I, when the Arab revolt was

brewing, British cartographers had no knowledge of the location of Medina, at the time the peninsula's largest city. And the map that Bertram Thomas drew for his book *Arabia Felix* would be the best available for close to forty years.

4. "the way they cut their hair . . .," Aubrey de Selincourt, trans., *Herodotus: The Histories* (New York: Penguin Books, 1985), p. 206.

5. *where Alexander never set foot.* Alexander the Great considered adding Arabia to his conquests, but he died in Babylon in 323 B.C. on the eve of his planned campaign. The fact that he wanted to invade the peninsula — and even make it his royal abode — is evidence that *something* in Arabia was of great value.

6. "The inhabitants of that place said . . .," Albert M. Wolohojian, trans., *The Romance of Alexander the Great by Pseudo-Callisthenes* (New York: Columbia University Press, 1969), pp. 112–15.

7. "Why do you tread this earth . . .," Wolohojian, *Romance of Alexander the Great*, p. 116.

8. 71° × 23° . . . 73° × 16°. Though the principle is the same, Ptolemy's latitudes and longitudes are not the same as those on modern maps, on which the 0° prime meridian passes through the British Royal Observatory at Greenwich. Ptolemy chose the Fortunate Islands in the Atlantic for his prime meridian, because they were considered the far edge of the habitable world. The scale of his system differs from ours as well.

4. The Flight of the Challenger

1. *its inherent distortions* . . . In laying out his maps, Ptolemy miscalculated the circumference of the earth. To make things fit, he took to squeezing empty spaces — such as the hinterlands of Arabia.

2. "It's like rotating your house . . .," Elachi quoted by Ronald Blom, Oct. 1984.

3. "the loss of viewing time . . .," *Time*, Oct. 22, 1984, p. 72.

4. "Have you not heard . . .," N. J. Dawood, trans., *The Koran* (New York: Penguin Books, 1981), p. 25.

5. *the desert of al-Ahqaf.* The Koran's passing mention of this region is a valuable geographic clue, for in post-Koranic accounts and even maps, al-Ahqaf (which has been taken to mean "wind-curved sand dunes") is located more or less where Bertram Thomas found his road to Ubar and where we hoped for success with the space shuttle's radar.

6. *the proper names "Iram" and "'Ad."* The very earliest evidence of the city of Iram and its People of 'Ad may be hidden in the word *Adramitae*, a southern Arabian tribe mentioned by Greek geographers. The name appears as well on Ptolemy's ever-helpful map of Arabia. Breaking the word apart, *Adrami-* could stand for "'Ad-i-Iram"and the suffix *-tae* means "tribe." The Adramitae, then, would be the People of 'Ad and Iram.

7. "Roast flesh, the glow of fiery wine . . .," Charles J. Lyall, *Ancient Arabian Poetry* (London: Williams & Norgate, 1930), p. 64.

8. "of ill omen . . .," Lyall, *Ancient Arabian Poetry*, p. 113; "She [War] brought forth Distress . . .," William A. Clouston, ed., *Arabian Poetry for English Readers* (Glasgow: McLaren & Son, 1881), p. 34.

9. "Arrogant and unjust . . ." This and the following quotes are from Dawood, *Koran*, pp. 159–60, 129, 205.

10. "According to the tradition of the Arabs . . .," Johann Burckhardt, *Travels in Arabia*, vol. 2 (Beirut: Librairie du Liban, 1972), p. 274; "the 'Adites continued to abandon themselves . . .," L. Du Couret, *Life in the Desert: Or, Recollections of Travel in Asia and Africa* (New York: Mason Brothers, 1860), p. 271.

11. *where lakes once formed.* The ancient lakes of the Rub' al-Khali have been extensively studied by geologist Hal McClure. His findings are succinctly summarized in Arthur Clark, "Lakes of the Rub' al-Khali," *Aramco World* 40, no. 3 (May–June 1989).

12. "Iram . . . will be unearthed by ants . . .," Nabih Amin Faris, trans., *The Antiquities of South Arabia* (Princeton: University Press, 1936), p. 72. The prediction that Ubar would be unearthed by ants isn't as cryptic as it might appear. There are several classical accounts of ants bringing gold to the surface in India. Herodotus says: "There is found in this desert a kind of ant of great size, bigger than a fox, though not so big as a dog . . . These creatures as they burrow underground throw up the sand in heaps, just like our own ants throw up the earth, and they are very like ours in shape. The sand has a rich content of gold, and this it is that the Indians are after when they make their expeditions into the desert" (de Selincourt, *Herodotus: The Histories*, p. 246). Could this ant be a lizard or small burrowing animal?

13. "Whoever shall find and enter Ubar . . .," David T. Rice, *The Illustrations of the "World History" of Rashid al-Din* (Edinburgh: University of Edinburgh Press, 1981), p. 42.

5. The Search Continues

1. "The bedu tell of such places . . .," Ranulph Fiennes, *Where Soldiers Fear to Tread* (London: Hodder & Stoughton, 1975), pp. 195–96.

2. *frankincense . . . exported by sea.* From Dhofar, incense was floated in animal-skin boats down the coast to the port of Qana (west of Mukalla in today's Yemen). There it was consigned to camel caravans bound across Arabia to the great caravansary of Petra (in today's Jordan).

6. The Inscription of the Crows

1. *converged at the well of Shisur.* Curiously, the old incense caravan route to and past Shisur is accurately marked on a map in

the classic 1911 edition of the *Encyclopaedia Britannica*. Where this information came from is a mystery; at the time, no westerner is known to have penetrated the region.

2. "And we hunted the game . . .," Rev. Charles Forster, *The Historical Geography of Arabia* (London: Duncan & Malcolm, 1844), pp. 90–93.

7. The Rawi's Tale

1. *rawis' tales of Iram/Ubar* . . . In the centuries after the Koran recounted the grief that befell the People of 'Ad, two competing story lines evolved. On one hand there is the tale of how Ubar's mighty king had a fabulous-beyond-belief city built in his absence, only to have it destroyed by God at the moment he and his retinue came in sight of it. In an alternate version, the city has long been inhabited and is known for its idolatry and dissolution. Ubar's king is warned by the Prophet Hud that disaster is imminent unless the People of 'Ad forsake their evil ways. Hud is ignored; the city is destroyed. This is the scenario of the excerpt in the text, which is from Part 12 of *The Tales of the Prophets of al-Kisa'i*, translated by W. M. Thackston, Jr. (Boston: Twayne Publishers, 1978), pp. 109–17.

2. "Oh my people . . . worship god . . ." To give *Tales of the Prophets* a sense of authenticity and a dash of piety, al-Kisa'i's direct quotes from the Koran were set off with the equivalent of italics.

3. "suddenly the earth opened . . .," Khairat al-Saleh, *Fabled Cities, Princes and Jinn* (London: Peter Stone, 1985), p. 45.

4. "Ubar is . . . the name of the land . . ." Medieval chroniclers who concur as to Iram/Ubar's location include Ibn Mujawir, Ibn Battuta, Ibn Ishaq, and al-Bedawi. Al-Himyari is quoted in Thomas, *Arabia Felix*, p. 161.

5. "They turned to dust . . .," al-Qadi Isma'il ibn Ali Al-Akoa,

"Nashwan Ibn Sa'id al-Himyari and the Spiritual, Religious and Political Conflicts of His Era," in Werner Daum, ed., *Yemen: 3000 Years of Art and Civilization in Arabia Felix* (Innsbruck: Pinguin-Verlag, 1988), p. 212.

8. Should You Eat Something That Talks to You?

1. "The Lord destroyed everything there . . .," Ferdinand Wusten-feld, ed., *Jacut's Geographisches Wörterbuch* (Leipzig: Bei F. A. Brockhaus, 1869), p. 897.
2. "Wabar is a vast piece of land . . .," Wustenfeld, *Jacut's Geographisches Wörterbuch*, pp. 866–68.

9. The City of Brass

1. "unwholesome literature . . .," quoted in Joseph Campbell, ed., *The Portable Arabian Nights* (New York: Viking Press, 1952), p. 1; "vulgar, insipid," quoted in Reynold A. Nicholson, A *Literary History of the Arabs* (Cambridge, Eng.: Cambridge University Press, 1930), p. 458; "The first who composed tales . . .," quoted in John Payne, *The Book of the Thousand Nights and One Night*, vol. 9 (London, 1884), p. 280.
2. *pre-Persian origin of the tales.* Frobenius hypothesized a common source for the Persian *Arabian Nights* and tales he collected from the Sudan, tales allegedly told by a slave named Far-li-mas, who hailed from the Arabian valley of the Hadramaut. Frobenius recalled that when he sailed the Red Sea in 1915, "the Arab seamen maintained, stoutly and firmly, that all the tales of the *Arabian Nights* had first been told in Hadramaut and from there had been diffused over the earth" (quoted in Joseph Campbell, *The Masks of God: Primitive Mythology* [New York: Penguin Books, 1986], p. 164). The subsequent diffusion of

the *Arabian Nights* to Persia may date to a Persian conquest of the Hadramaut, a little-known chapter of Arabian history.

3. "Allah blotted out the road . . .," Richard F. Burton, trans., *The Book of a Thousand Nights and a Night*, vol. 4 (London, 1885), p. 116. It is quite possible that the writer of this tale was familiar with an account (circa 1300) by Ibn Mujawir, a merchant of Baghdad, who wrote of an old, abandoned caravan road from Baghdad to southern Arabia. It was a direct route across the Rub' al-Khali, and it almost certainly would have passed through our search area. Was Ibn Mujawir describing our road to Ubar? If so, he advised against following it, for it was dangerous, abandoned for good reason. He wrote: "God is a witness that any bedouin who travels this road again has no one but himself to blame!" (Quoted in G. Rex Smith, "Ibn al-Mujawir on Dhofar and Socotra," in *Proceedings of the Eighteenth Seminar for Arabian Studies* [London: Seminar for Arabian Studies, 1985], pp. 84–85.)

4. "had been translated . . .," from "The Eldest Lady's Tale," in Burton, *Thousand Nights*, vol. 1, p. 165. Under the name "The Petrified City," this tale appears in Wil Clap's *Oriental Moralist* of 1797.

5. "When they reached the top . . ." This and the following quotes are from Burton, *Thousand Nights*, vol. 6, pp. 102, 114–15, 93, 119.

6. *an ancient language . . .* The language of the Dhofar Mountains is actually a cluster of four related tongues called the Hadara group. Shahri, believed to be the oldest, is described in Chapter 12.

10. The Singing Sands

1. "Wabar, it seemed . . ." This and the following quotes are from Josephine Tey, *The Singing Sands* (New York: Collier Books, 1988), pp. 140, 176, 205, 141.

11. *Reconnaissance*

1. "The plan is great," "the great scheme," "Asadum Tal'an . . .,"
   "The one-eyed . . . D E T E S T A B L E !" Jacqueline Pirenne, "The
   Incense Port of Moscha (Khor Rori) in Dhofar," *Journal of
   Oman Studies* 1 (Ministry of Information and Culture, Sultanate
   of Oman, 1975), pp. 82, 86, 89, 90.
2. *Andhur flint could have been traded* . . . With chemical finger-
   printing, Juri explained, the extent of Andhur's flint trade could
   accurately be charted. Finding Andhur flint in sites to the north
   of the Rub' al-Khali would confirm the long-range reach — and
   trading importance — of our Ubar road.

12. *The Edge of the Known World*

1. "half a day's journey . . ." Ibn Battuta quoted in Philip Ward,
   Travels in Oman (New York: Oleander Press, 1978), p. 503.
2. *the well of the Oracle of 'Ad.* We were not the first to reconnoiter
   the well. The intrepid husband and wife team of Theodore and
   Mabel Bent had been here in 1895, Bertram Thomas in 1929,
   and Wendell Phillips in 1953. None had known quite what to
   make of it. See Bent and Bent, *Southern Arabia* (London: Smith,
   Elder, 1900); Thomas, *Arabia Felix*; and Frank P. Albright, *The
   American Archaeological Expedition in Dhofar, Oman* (Washington: American Foundation for the Study of Man, 1982).
3. "guarded by flying serpents," Selincourt, *Herodotus: The Histories*, p. 249; "in the most fragrant forests . . .," C. H. Oldfather,
   trans., *Diodorus of Sicily* (Cambridge: Harvard University Press,
   1979), p. 229; "sprang as high as the thigh . . .," Howard L. Jones,
   trans., *Strabo: Geography*, vol. 7 (Cambridge: Harvard University Press, 1995), p. 347.
4. "the language of birds." This remarkable language was reported
   by Theodore and Mabel Bent (*Southern Arabia*) in 1900. As

early as five thousand years ago, the Sumerians called an aboriginal tribe near the Persian Gulf "Lulubulu," an onomatopoeic word mimicking the song of birds. It's quite possible they could have been describing the same language, as it was spoken by the ancestors of today's Shahra.

5. *They were frankincense trees . . .* Frankincense trees — *Boswellia sacra* — are found elsewhere in Arabia and even in Africa. Though often impressive in size, none produce the pure, ethereally fragrant resin of the small, tortured trees of the Dhofar Mountains. Perhaps it is because the trees there grow in a unique microclimate: an elevation of 600–700 meters and seasons that alternate scorching sun with monsoon drizzle.

6. "No Latin writer . . .," "The district . . . is rendered inaccessible . . .," and "It is the people who originated the trade . . .," John Bostock and H. T. Riley, trans., *The Natural History of Pliny*, vol. 3 (London: Henry G. Bohn, 1855), pp. 124, 125.

7. "Look at this your sacrifice . . ." The Shahra's timeless chant of exorcism, first recorded by Bertram Thomas in 1930, was unchanged sixty years later.

8. *Here was a living link . . .* The Shahra also speak of al-Ahqaf, the Koran's location of our lost city. They consider al-Ahqaf to be not only the sands beyond their mountains (which we believed), but the mountains themselves. This made sense, for whoever built Ubar would have also held sway over the incense groves of the Dhofar Mountains.

13. The Vale of Remembrance

1. *triliths were memorials . . .* It's doubtful that triliths marked actual burials, for some were set on exposed bedrock, where interment would have been impossible. A more reasonable explanation would be that they honored dead laid to rest elsewhere, as in the cave of the skulls we had visited.

2. "Whenever a traveler stopped . . .," Nabih Amin Faris, trans., *The Book of Idols* (Princeton: Princeton University Press, 1952), pp. 28–29.

3. "the secret of God in the universe . . .," Ali Al-Shari'at, *Hajj* (Tehran: Laleh-Baktiar, 1988), p. 48.

4. "went so far as to pay divine worship . . .," George Sale, trans., "Preliminary Discourse," *The Koran* (London: Thomas Tegg & Son, 1838), p. 15.

14. The Empty Quarter

1. "This wilderness . . . stretches away . . .," S. B. Miles, *The Countries and Tribes of the Persian Gulf* (London: Frank Cass, 1919), p. 386.

2. *Ron guessed that the satellites . . .* We later learned that the satellite navigation system had been more or less shut down for realignment — precisely when we planned to rely on it for our journey into the Rub' al-Khali.

3. "Only a fool will brave the desert sun . . .," O'Shea, *Sand Kings of Oman*, p. 187.

4. *The first half dozen he threw . . .* "AFR" is an informal archaeological designation for worthless: "A" stands for "another," and "R" for "rock."

5. "a creature of night to signify the days . . .," cited in John Gray, *Near Eastern Mythology* (New York: Peter Bedrick Books, 1982), p. 37.

16. City of Towers

1. "Arrogant and unjust were the men of 'Ad . . ." and the following quotes are from Dawood, *Koran*, pp. 159, 25, 113, 129, and 138.

2. *Our Christmas tree.* Out of deference to our Islamic host country, we had anticipated a low-key Christmas and brought with us

but a single tape. To our surprise, the Omanis loved the holiday. When we drove to the coast to collect Juri's students, it was to the strains (on the radio, in the hotel, everywhere) of familiar carols. We heartily sang along with "We Three Kings of Orient are . . .," for we were in that very Orient, the land of frankincense and myrrh.

3. *It might have been used . . . to process frankincense.* How frankincense was processed is unclear. Its crystals may have been compacted for shipment, or a refining process may have enhanced its aroma. From the historian Pliny we do know that frankincense was processed at the far end of the Incense Road. He writes: "At Alexandria . . . where the frankincense is worked up for sale, good heavens! no vigilance is sufficient to guard the factories. A seal is put upon the workmen's aprons, they have to wear a mask or a net with a close mesh on their heads, and they are stripped naked before they are allowed to leave work" (Bostock and Riley, *Natural History of Pliny*, 3: 127).

4. *Where were the columns?* After the expedition I discovered that in pre-Islamic poetry (the literature closest to the era of Ubar) the word for pillar is not عماد *imad* but دوّار *dawwar*. Appearing only once in the Koran, the word *imad* appears to be a southern Babylonian loan word derived from a root meaning "to make stand, to erect" — and can describe anything from tent poles to pillars to towers.

5. *a sprawling oasis.* Mabrook recalled that as late as the 1920s, his grandfather remembered a dense "forest" of brush and dwarf trees in the outlying area known as Hailat Shisur. And in the 1930s Bertram Thomas wrote, "I have heard that in the surrounding desert plain are still to be seen shadowed furrowings as though once it had known the plough" (Thomas, *Arabia Felix*, p. 137).

6. *a six-pointed star.* Was our chess king's star a star of David? I later

learned that the six-pointed star, though linked to Judaism from the 1500s on, may have been no more than a popular (and secular) design motif in the Middle East before then. Yet six-pointed stars have been discovered at an early synagogue at Capernaum in ancient Palestine, on the third-century tombstone of a certain Leon ben (son of) David, and in Jewish catacombs near Rome. And now on a chess king in the Arabian desert.

7. *back to their mountain retreat.* Golden grave goods may yet come to light in the Vale of Remembrance, though it is doubtful. The extent of grave robbing in southern Arabia is reflected in the fact that a major function of the god of the morning star was to avenge desecrators of the dead.

8. *inscription that included the word*)Πᕼ. As several experts assured us, in the more than ten thousand known southern Arabian inscriptions, the word)Πᕼ was nowhere to be found. But then I happened on it in an inscription found at an Arabian colony in ancient Ethiopia. Jacqueline Pirenne equates the word with *Abiru,* meaning "Hebrew." This could be evidence of a Jewish association with Ubar (an association already present in the figure of the prophet Hud, "He of the Jews"). Or this could be a wishful translation (Father Jamme thinks it is), and *Ubar* could instead be derived from the Semitic root for either "place of passage" or "camel hair tent."

18. Seasons in the Land of Frankincense

1. *a Mesopotamian-Persian sphere of influence.* A link between Ubar and Mesopotamia tallies with a fragment of myth in which "the 'Adites quarreled with the children of Ham and left Babylon. They peopled a district in southern Arabia contiguous to 'Umman, Yaman, and Hadramaut. There they built palaces, erected temples, and worshipped deities as stars."

2. *eastern versus western Arabia.* An Arabian east versus west map can be drawn with archaeological evidence, admittedly sketchy, and with a brand-new cultural resource: genetic mapping. The division on the map on page 209 is based on the mean strength of the genes ESD*1 and GC*1F.

3. *a temple as well as an administrative center.* In *Ancient Yemen* (Oxford: Oxford University Press, 1995), Andrey Korotoyev analyzes a settlement pattern in which a *hagar* was a *dual* religious and political center for a *"sha'b,"* a surrounding territory of several dozen square kilometers. Ubar would have been a hagar.

4. "Show us our Christ, alas!" and "Whereupon, after a terrible storm . . .," Sale, "Preliminary Discourse," *Koran,* p. 16.

5. *traditions of desert life.* The renowned Cambridge Arabist Robert Serjeant found southern Arabia ideal for the concept of "Interpretation of the Antique by Reference to the Present" (Serjeant, *South Arabian Hunt* [London: Luzac, 1983], p. 80).

19. Older Than 'Ad

1. *At these sites . . .* Archaeologically, the sites near Shisur contemporary with the Rub' al-Khali's lakes are Upper Paleolithic (40,000–100,000 years before the present). Juris Zarins homed in on some forty small settlements from this era by plotting the courses of late Pleistocene rivers found on space images, then methodically searching their banks.

2. *the rains withdrew.* The onset of hyperaridity was caused by a phenomenon called "Milankovitch forcing," in which Earth wobbled slightly in its orbit around the sun. This precipitated a global climactic change that to a large extent initiated the desertification of Arabia, Africa, India, and Australia.

3. *retreating to the north . . .* It has long been argued — and counterargued — that the Semitic populations of the Middle East

arose from the deserts of Arabia. A migration north twenty thousand years ago is how and when this could have happened. Though the date of this migration is far earlier than biblical scholars would like, the idea has an appealing fit. It has recently been championed by geologist Hal McClure.

4. *waiting hunters would rise up* . . . The outline of Shisur's impressive Neolithic animal trap was photographed — quite unintentionally — by an Omani military overflight in the late 1970s. In 1990, Shisur's new village obliterated all traces of it.

5. "smelled the sweet savor . . .," Alexander Heidel, *The Gilgamesh Epic and Old Testament Parallels* (Chicago: University of Chicago Press, 1963), p. 87; "its resin was considered . . .," Walter W. Müller, "Notes on the Use of Frankincense in South Arabia," in *Proceedings of the Ninth Seminar for Arabian Studies* (London: Seminar for Arabian Studies, 1976), p. 131.

20. The Incense Trade

1. "the rising of the Dog Star . . .," Bostock and Riley, *Natural History of Pliny*, vol. 3, pp. 126–27.

2. *the surrounding oasis.* An ancient oasis appears to have extended east from Shisur along a fault line that tapped an aquifer charged by the runoff from the Dhofar Mountains. To this day, the wadi overlying this fault is called Umm al-Hait, the Mother of Life.

3. "The fairness of beautiful girls . . .," Thomas, *Alarms and Excursions*, p. 288.

4. "Thou shalt cast incense . . .," Master of Belhaven (A. Hamilton), *The Kingdom of Melchior* (London: John Murray, 1949), pp. 21, 20; "A stairway to the sky . . .," Raymond O. Faulkner, *The Ancient Egyptian Pyramid Texts* (Oxford: Oxford University Press, 1969), p. 76.

5. "called sacred and . . . not allowed . . .," Bostock and Riley, *Natural History of Pliny*, vol. 3, p. 125.

6. "The whole city now is conceived . . .," Joseph Campbell, "The Hieratic City State," *Parabola* 18, no. 4 (Nov. 1993), pp. 41–43.

7. *The language of the 'Adites . . .* Though only a two-letter fragment of the 'Ad script has surfaced at Ubar, inscriptions abound in the Dhofar Mountains. It appears to have preceded not only other languages of southern Arabia, but also Hebrew, which has nine fewer sounds, and Arabic, which has eight fewer.

8. "broken heads . . . and to bind bloody wounds . . .," Oldfather, *Diodorus of Sicily*, p. 45.

9. "It is the luxury of man . . .," Bostock and Riley, *Natural History of Pliny*, vol. 3, p. 127. As a measure of the value of frankincense, there are records of the denarii paid for a measure in the markets of Rome. To translate its cost into modern terms, the Smithsonian's Gus Van Beek worked out the formula that a pound of frankincense was worth between 2.5 and 5 percent of the minimum annual urban cost of living. In 1990 dollars, that would be over $1,000 a pound.

21. Khuljan's City

1. *summer's night in 350 B.C.* This year could have been as early as 410 or as late as 290 B.C. The earliest carbon-14 date associated with Ubar's New City is 350 B.C. plus or minus sixty years.

2. "After the sun has set . . .," Thomas, *Arabia Felix*, pp. 52, 290.

3. *a fence woven of gnarled branches . . .* Duwwar construction is still used by the Shahra of the Dhofar Mountains. Its use at Ubar would explain why there is no "meltdown" from dissolved mud brick walls.

4. "To thee from Babylon we made our way . . .," Faris, *Antiquities of South Arabia*, p. 30.

5. *a large plastered basin . . .* Water installations — including fountains and sheets of water one walked through — were an important feature of Arabian temples. To ensure a fresh water supply,

there may have been a rock-cut passage between Ubar's temple and the spring. Our Shisur friend Baheet recalled that as a boy, he found and squeezed through such a passage that had since collapsed.

6. *the kahin shuffled the arrows* . . . Divination by arrows is called *istqam* in Arabic, "rhabdomancy" in English. The rite may have an echo in the Bible when Yahweh, through his soothsayer Gad, speaks to David, saying: "I offer you three things; choose one of them for me to do to you" (2 Samuel 24:12). Similarly, in legend, the fate of the Ubarites is sealed as they choose one of three clouds.

7. *Once . . . the religion of the 'Ad may have been more meaningful* . . . As historian Karen Armstrong has noted, "In Arabia the original symbolic significance of the old gods had been lost during the nomadic period and Arab religion had no developed mythology to express this pagan insight" (Armstrong, *Muhammad: A Biography of the Prophet* [San Francisco: Harper San Francisco, 1992], p. 98).

8. "Were it not for her whose wily charms . . .," Faris, *Antiquities of South Arabia*, p. 29.

9. "Roast flesh, the glow of fiery wine . . .," Lyall, *Ancient Arabian Poetry*, p. 64.

10. *to Eriyot, his royal city*. Eriyot may have been Ain Humran. But there is also reason to believe that Eriyot may have been obliterated by the construction of the sultan of Oman's Robat palace in Salalah.

22. City of Good and Evil

1. "A singular thing too . . .," Bostock and Riley, *Natural History of Pliny*, vol. 2, p. 91 (italics added). Elsewhere (vol. 3, p. 135) Pliny notes an interesting exception to his "they purchase nothing in

return" statement. He tells us that "in Arabia there is a surprising demand for foreign scents, which are imported from abroad; so soon are mortals sated with what they have of their own, and so covetous are they of what belongs to others."

2. *Ubar continued to prosper.* The coastal incursion by the kingdom of the Hadramaut could actually have been a boon for Ubar, as the 'Adites chose to ship more and more of their incense overland rather than sell it to the Hadramis garrisoned at the port of Sumhuram.

3. *a Jewish king sat on the throne . . .* Yusuf As'ar Yath'ar, lord of Dhu Nuwas, was "King of all Tribes." He ruled over the Himyar, a people who conquered the old city-states of Ma'in, Qataban, and Saba.

4. "which Shaddad ibn 'Ad built . . .," Thackston, *Tales of the Prophets*, p. 126.

5. *the vocabulary of pre-Islamic Arabians . . .* The glossary of Father Jamme's *Inscriptions at Mahram Bilqis* provides an excellent overview of what was on the minds of the southern Arabians from approximately 750 B.C. to 450 A.D.

6. "Brothers are held in higher honor . . .," Jones, *Geography of Strabo*, pp. 365–66.

7. "when an Arab had a daughter born . . .," cited in Sale, *Koran*, p. 94.

23. Sons and Thrones Are Destroyed

1. *the story became part and parcel of Jewish folklore.* The historian al-Tabari reports that *before* the time of Muhammad the Jews of western Arabia threatened their enemies: "We shall kill you as 'Ad and Iram were killed" (Edshan Yar-Shater, ed., *The History of al-Tabari*, vol. 6 [Albany: State University of New York Press, 1989], pp. 124–25).

2. "We, the dwellers in this palace . . ." and "I, Shaddad ben 'Ad, ruled . . .," Louis Ginzberg, *The Legends of the Bible* (Philadelphia: Jewish Publication Society of America, 1913), pp. 590, 571.

3. "Whoever doth read this writing . . .," Angelo S. Rappoport, *Ancient Israel: Myths and Legends* (New York: Bonanza Books, 1987), vol. 3, p. 106.

4. "As old as 'Ad . . .," Thomas P. Hughes, A *Dictionary of Islam* (Lahore: Premier Book House, 1986), p. 18; "Roast flesh, the glow of fiery wine . . .," Lyall, *Ancient Arabian Poetry*, p. 64; "And ninety concubines . . .," Philby, *Empty Quarter*, p. 157; "O delegation of drunks . . .," Thackston, *Tales of the Prophets*, pp. 114, 116; "Wealth, easy lot . . .," Lyall, *Ancient Arabian Poetry*, p. 64; "An ignominious punishment . . .," Dawood, *Koran*, pp. 128–29; "Sons and thrones are destroyed . . .," Thackston, *Tales of the Prophets*, p. 116; "Now all is gone . . .," Philby, *Empty Quarter*, p. 157; "Checkmate . . . It was a great city . . .," Thomas, *Arabia Felix*, p. 161; "At the end of life . . .," Edward Rice, *Captain Sir Francis Burton* (New York: Charles Scribner's Sons, 1990), p. 440.

Epilogue: Hud's Tomb

1. *patriarchs and prophets holy to Islam.* This tier of mythological landscape includes figures that Islam shares with Judaism and Christianity. In the seaside town of Salalah, the tomb of Nebi Umran, father of the Virgin Mary, is venerated; in the mountains above Salalah, pilgrims leave offerings of incense and flowers at Job's tomb; in the desert beyond, at a spring called Mudhai, the bedouin will show you where Moses hit a rock seven times with his staff, and water magically flowed. It still does.

2. *perhaps original tomb . . .* The medieval traveler Ibn Battuta reported that in Dhofar there was a building containing a grave on

which is inscribed "This is the grave of Hud ibn 'Abir. God bless and save him." Here was a choice bit of evidence linking Hud to Ubar ('Abir).

3. *a perhaps less authentic . . . Hud's tomb . . .* His second tomb, in the Valley of the Hadramaut, came into prominence at the earliest in the 900s, when it was "rediscovered" under questionable circumstances by a pair out of the *Arabian Nights*, a saintly descendant of Muhammad and a camel-driving scoundrel. For the record, there are even more Hud's tombs: a third is near the well of Zamzam in Mecca; a fourth is outside the town of Salt in Jordan; and a fifth is in the south wall of the great mosque of Damascus.

4. *the world's first skyscrapers.* Some of Shibam's skyscrapers date to as early as the 900s. Well before then, as revealed by recent archaeology, pre-Islamic buildings in the Hadramaut rose as high as seven stories. (See Jacques Seigne, "Le Château Royal de Shabwa: Architecture, Techniques de Construction et Restitutions," *Syria* 68 [1991].)

5. "You camel men, go . . ." and "O Khon, no girl in Khon . . .," cited in R. B. Serjeant, "Hud and Other Pre-Islamic Prophets of Hadramawt," *Le Museon* 57 (1954), p. 25.

6. "God curse you, infidel woman," Serjeant, "Hud and Other Pre-Islamic Prophets," p. 29; "invoke peace on all the prophets . . .," Harold Ingrams, *Arabia and the Isles* (London: John Murray, 1942), p. 215.

7. "As I went in, I saw . . .," Faris, *Antiquities of South Arabia*, pp. 79–80. For another version of this, see al-Kisai's tale of the prophet Hud in Chapter 7, page 87.

8. "So sacred that a stick . . .," Ingrams, *Arabia and the Isles*, p. 216.

9. *a sulfurous portal to the underworld.* The account of one al-Qazwini in 1250 tells us: "The bit of earth most hated by Allah is Wadi Barhut, in which there is a well filled with evil-smelling,

black water, wherein go the souls of the unbelievers." He also cites a Hadrami belief that "whenever we notice a foul smell in the neighborhood of Barhut, then, later on, we are informed of the death of one of the most prominent among the unbelievers" (Ferdinand Wustenfeld, ed., 'Adjaib al-Makhluqat, vol. 1 [Gottingen, 1849], p. 198).

Bibliography

Albright, Frank P. *The American Archaeological Expedition in Dhofar, Oman*. Washington: American Foundation for the Study
   of Man, 1982.

Ali, S. M. *Arab Geography*. Aligarh, India: Institute of Islamic Studies, Muslim University, 1960.

Armstrong, Karen. *Muhammad: A Biography of the Prophet*. San
   Francisco: Harper San Francisco, 1992.

———. *A History of God*. New York: Alfred A. Knopf, 1994.

Bagrow, Leo. *History of Cartography*. Chicago: Precedent, 1985.

Beeston, A. F. L. "Functional Significance of the Old South Arabia
   'Town,'" in *Proceedings of the Seminar for Arabian Studies*.
   London: Seminar for Arabian Studies, 1971.

———. "The Settlement at Khor Rori," *Journal of Oman Studies* 2
   (1976).

———. "The Religions of Pre-Islamic Yemen" and "Judaism and
   Christianity in Pre-Islamic Yemen," in Joseph Chelhod, ed.,
   L'Arabie du Sud, Histoire et Civilization, vol. 1: *Le Peuple
   Yemenite et Ses Racines*. Paris: Editions G.-P. Maisonneuve
   et Larose, 1984.

Beeston, A. F. L., T. M. Johnstone, R. B. Serjeant, and G. R. Smith.
   Arabic Literature to the End of the Umayyad Period. Cambridge, Eng.: Cambridge University Press, 1983.

Bent, Peter. *Far Arabia: Explorers of the Myth*. London: Weidenfeld & Nicolson, 1977.

Bent, Theodore, and Mrs. Theodore Bent. *Southern Arabia*. London: Smith, Elder, 1900.

Bidez, Joseph, ed. *Philostorgius Kirchengeschichte*. Berlin: Akademie-Verlag, 1981.

Biella, Joan Copeland. *Dictionary of Old South Arabic*. Cambridge, Mass.: Scholars Press, 1982.

Bilkadi, Zayn. "The Wabar Meteorite," *Aramco World Magazine* 37, no. 6 (Nov.–Dec. 1986).

Blom, Ronald. "Space Technology and the Discovery of Ubar," *P.O.B.* 17, no. 6 (Aug.–Sept. 1992).

————. "The Discovery of Ubar: Use of Space-Based Image Data in the Search for Ubar." Paper presented to UCLA University Course X 401 (Oct. 1992).

Blom, Ronald, and Charles Elachi. "Spaceborne and Airborne Imaging Radar of Sand Dunes," *Journal of Geophysical Research* 86, no. B4 (Apr. 10, 1981).

Bostock, John, and H. T. Riley, trans. *The Natural History of Pliny*. London: Henry G. Bohn, 1885.

Bowen, Richard L., Jr., and Frank P. Albright, eds. *Archaeological Discoveries in South Arabia*. Baltimore: Johns Hopkins University Press, 1958.

Bravmann, H. H. *The Spiritual Background of Early Islam*. Leiden: E. J. Brill, 1972.

Breton, J.-F. "Religious Architecture in Ancient Hadramawt," in *Proceedings of the Thirteenth Seminar for Arabian Studies*. London: Seminar for Arabian Studies, 1980.

Brice, William C. "The Construction of Ptolemy's Map of South Arabia," in *Proceedings of the Seminar for Arabian Studies*, vol. 4. London: Seminar for Arabian Studies, 1974.

Budge, Ernest A. Wallis. *The Alexander Book in Ethiopia*. London: Oxford University Press, 1933.

Bulliet, Richard W. *The Camel and the Wheel.* Cambridge, Mass.: Harvard University Press, 1975.

Burton, Richard F., trans. *The Book of a Thousand Nights and a Night.* London: Burton Club, for Private Subscribers Only, 1885.

Campbell, Joseph. *The Masks of God.* Vol. 1: *Primitive Mythology;* vol. 2: *Occidental Mythology.* New York: Penguin Books, 1986.

————. "The Hieratic City State," *Parabola* 18, no. 4 (Nov. 1993).

————, ed. *The Portable Arabian Nights.* New York: Viking Press, 1952.

Casson, Lionel, trans. *The Periplus Maris Erythraei.* Princeton: Princeton University Press, 1989.

Caton-Thompson, G., and E. W. Gardner. "Climate, Irrigation, and Early Man in the Hadhramaut," *Geographical Journal* 93, no. 1 (Jan. 1939).

————. *The Tombs and Moon Temple at Hureidha (Hadhramaut).* Oxford, Eng.: Oxford University Press, 1944.

Cavalli-Sforza, L. Luca, Paolo Menozzi, and Alberto Piazzi. *The History and Biography of Human Genes.* Princeton: Princeton University Press, 1994.

Chatty, Dawn. "The Bedouin of Central Oman," *Journal of Oman Studies* 6, pt. 1 (1993).

Clark, Anthony. *Seeing Beneath the Soil: Prospecting Methods in Archaeology.* London: B. T. Batsford, 1990.

Clark, Arthur. "Lakes of the Rub' al-Khali," *Aramco World Magazine* 40, no. 3 (May–June 1989).

Cleuziou, Serge. "Oman Peninsula in the Early Second Millennium B.C.," in H. Hartel, ed., *South Asian Archaeology 1979.* Berlin: D. Reimer Verlag, 1981.

————. "Hili and the Beginning of Oasis Life in Eastern Arabia," in *Proceedings of the Fifteenth Seminar for Arabian Studies.* London: Seminar for Arabian Studies, 1982.

Cleveland, Ray L. "The 1960 American Archaeological Expedition to Dhofar," *Bulletin of the American Schools of Oriental Research* 159 (Oct. 1960).

Cohen, Shaye J. D., and Ernest S. Frerichs, eds. *Diasporas in Antiquity.* Atlanta: Scholars Press, 1981.

Coon, Carleton S. "Southern Arabia: A Problem for the Future," in *Smithsonian Report for 1944.* Washington: U.S. Government Printing Office, 1945.

Cooper, Rev. Mr. J., trans. *The Oriental Moralist or the Beauties of the Arabian Nights Entertainments.* Dover, N.H.: Printed by Samuel Bragg, Jr., for Wm. T. Clap, Boston, 1797.

Crichton, Andrew. *History of Arabia, Ancient and Modern.* Edinburgh: Oliver & Boyd, 1834.

Crippen, Robert E. "The Regression Intersection Method of Adjusting Image Data for Band Rationing," *International Journal of Remote Sensing* 8, no. 2 (1987).

Crippen, Robert E., Ronald Blom, and Jan R. Hayada. "Directed Band Rationing for the Retention of Perceptually Independent Topographic Expression in Chromaticity-Enhanced Imagery," *International Journal of Remote Sensing* 9, no. 4 (1988).

D'Anville, Jean Baptiste. *Compendium of Ancient Geography.* London: J. Faulder, 1810.

Daum, Werner, ed. *Yemen: 3000 Years of Art and Civilization in Arabia Felix.* Innsbruck: Pinguin-Verlag, 1988.

Dawood, N. J., trans. *The Koran.* New York: Penguin Books, 1981.

Debevoise, Neilson C. *A Political History of Parthia.* New York: Greenwood Press, 1968.

De Gaury, Gerald, and H. V. F. Winstone. *Spirit of the East.* London: Quartet Books, 1979.

Dickson, H. R. P. *The Arab of the Desert.* London: George Allen & Unwin, 1959.

Doe, Brian. *Southern Arabia*. London: Thames & Hudson, 1971.

———. *Monuments of South Arabia*. London: Falcon-Oleander, 1983.

Dostal, Walther. *Die Beduinen in Sudarabia*. Vienna: Verlag F. Berger, Sohne Horn, 1967.

Du Couret, Colonel L. *Life in the Desert: Or, Recollections of Travel in Asia and Africa*. New York: Mason Brothers, 1860.

Easton, Gai. "Light from the Center," *Parabola* 18, no. 4 (Nov. 1993).

Eberhart, Jonathan. "Radar from Space: The Sightseeing Plans of SIR-B," *Science News* 126, no. 12 (Sept. 22, 1984).

Edens, Christopher. "The Rub' al-Khali Neolithic Revisited: The View from Nadqan," in D. Potts, ed., *Araby the Blest: Studies in Arabian Archaeology*. Copenhagen: Museum Tuscalanum Press, 1988.

Elachi, Charles, and Ronald Blom. "Seeing through Sand," *Planetary Report* 3, no. 5 (Sept.–Oct. 1983).

Elachi, Charles, G. Schaber, and L. Roth. "Spaceborne Radar Subsurface Imaging in Hyperarid Regions," *IEEE Transactions on Geoscience and Remote Sensing*, GE-22 (1984).

Eliade, Mircea. *A History of Religious Ideas*. Chicago: University of Chicago Press, 1978.

Faris, Nabih Amin. *The Antiquities of South Arabia*. Translation of Abu Muhammad al-Hamdani, *The Eighth Book of al-Iklil*. Princeton: Princeton University Press, 1936.

———. *The Book of Idols*. Translation of Hisham ibn al-Kalbi, *Kitab al-Asnam*. Princeton: Princeton University Press, 1952.

Fiennes, Ranulph. *Where Soldiers Fear to Tread*. London: Hodder & Stoughton, 1975.

Forster, Rev. Charles. *The Historical Geography of Arabia*. London: Duncan & Malcolm, 1844.

Freedman, David Noel, ed. Entries on incense, frankincense, Gad, prehistory of Arabia, religion of South Arabia, and Zoroastrianism, *The Anchor Bible Dictionary.* New York: Doubleday, 1992.

Gerhardt, Mia I. *The Art of Storytelling.* Leiden: E. J. Brill, 1963.

al-Ghazali, Abu Hamid. *The Remembrance of Death and the Afterlife.* Translated by T. J. Winter. Cambridge, Eng.: Islamic Texts Society, 1989.

Gibb, H. A. R., trans. *The Travels of Ibn Battuta.* Cambridge: Cambridge University Press, 1956.

Gibb, H. A. R. et al., eds. Entries on 'Ad, Alf Layla wa-Layla, Badw, Barhut, djinn, al-Hamdani, Hud, Iram, Lukman, Mahra, Nasara, Wabar, Yaqut, and Zafar, *The Encyclopaedia of Islam.* Leiden: E. J. Brill, 1979.

Ginzberg, Louis. *The Legends of the Bible.* Philadelphia: Jewish Publication Society of America, 1913.

Gray, John. *Near Eastern Mythology.* New York: Peter Bedrick Books, 1982.

Grohmann, Adolph. *Arabien.* Munich: C. H. Beck'sche Verlagsbuchhandlung, 1963.

Groom, Nigel. *Frankincense and Myrrh.* London: Longman, 1981.

————. "Oman and the Emirates in Ptolemy's Map," *Arabian Archaeology and Epigraphy* 5 (1994).

Guest, Rhuvon. "Zufar in the Middle Ages," *Islamic Culture: The Hyderabad Quarterly Review* 61, no. 3 (July 1992).

Gunther, Robert T. *The Greek Herbal of Dioscorides.* Oxford: Oxford University Press, 1934.

Haack, H. *Stieler's Atlas of Modern Geography.* Gotha, Germany: Justus Perthes, 1932/34.

Hamarneh, Saleh. "The Iconography of Idols and Its Significance for the Arabs before Islam," in Fawzi Zayadine, ed., *Petra and the Caravan Cities.* Amman, Jordan: Department of Antiquities, 1990.

Hamblin, Dora J. "Treasures of the Sands," *Smithsonian* 14, no. 6 (Sept. 1983).

Harding, G. Lankester. *Archaeology in the Aden Protectorates*. London: Her Majesty's Stationery Office, 1964.

Hawley, Donald. *Oman and Its Renaissance*. London: Stacey International, 1990.

Hennig, Richard. *Terrae Incognitae*. Leiden: E. J. Brill, 1944.

Hettner, Alfred. *Geographische Zeitschrift*. Leipzig: B. G. Teubner, 1934.

Hitti, Philip. *History of the Arabs*. New York: St. Martin's Press, 1964.

Hogarth, David G. *The Penetration of Arabia*. New York: Frederick A. Stokes, 1904.

Holt, P. M., Ann K. S. Lambton, and Bernard Lewis, eds. *The Cambridge History of Islam*. Cambridge, Eng.: Cambridge University Press, 1977.

Horovitz, Josef. *Koranische Untersuchungen*. Berlin: Walter De Gruyter, 1926.

Hughes, Thomas P. *A Dictionary of Islam*. Lahore, Pakistan: Premier Book House, 1986.

Ingrams, Harold. *Arabia and the Isles*. London: John Murray, 1942.

Jamme, Albert. *Sabaean Inscriptions from Mahram Bilqis*. Baltimore: Johns Hopkins University Press, 1962.

———. "Two Tham Inscriptions from Northern Dhofar, Oman, JaT 97 and 98," *Miscellanées d'ancien Arabe* 17 (Oct. 17, 1988).

Janzen, Jorg. *Nomads in the Sultanate of Oman: Tradition and Development in Dhofar*. Boulder, Colo.: Westview Press, 1986.

Johnstone, T. M. "Folk-tales and Folk-lore of Dhofar," *Journal of Oman Studies* 6, 1 (1980).

Jones, Alexander, ed. *The Jerusalem Bible*. Garden City, N.Y.: Doubleday, 1966.

Jones, Horace L., trans. *The Geography of Strabo*. Cambridge, Mass.: Harvard University Press, 1983.

Joukowsky, Martha. *A Complete Field Manual of Archaeology*. Englewood Cliffs, N.J.: Prentice-Hall, 1980.

Joyce, Christopher. "Archaeology Takes to the Skies," *New Scientist,* Jan. 25, 1992.

Kangas, B. "Introduction to Ethnomedicine: Examples from Yemen," *Bulletin of American Institute of Yemeni Studies* 35 (Summer–Fall 1994).

Keane, A. H. *The Gold of Ophir.* London: Edward Stanford, 1901.

Khan, Muhammad Muhsin. *The Translation of the Meanings of Sahih Al-Bukhari.* Chicago: Kazi Publications, 1979.

Knappert, J. "The Qisasu'l-Anbiya'i as Moralistic Stories," in *Proceedings of the Ninth Seminar for Arabian Studies.* London: Seminar for Arabian Studies, 1976.

Korotoyev, Andrey. *Ancient Yemen.* Oxford: Oxford University Press, 1995.

Kropp, Manfred. *Die Geschichte der "reiner Araber" vom Stamme Qahtan.* Frankfurt: Verlag Peter Lang, 1982.

Lammens, Henri. *Le berceau de l'Islam.* Rome: Sumptibus Pontificii Institui Biblici, 1914.

Landberg, Le comte de. *Etudes sur les Dialectes de l'Arabie Meridionale.* Leiden: E. J. Brill, 1901.

Lewcock, Ronald B. *Wadi Hadramawt and the Walled City of Shibam.* Paris: UNESCO, 1986.

Lightfoot, Victoria, and Dale Lightfoot. "Revealing the Ancient World through High Technology," *Technology Review,* May–June, 1989.

Lissner, Ivar. *The Silent Past.* New York: G. P. Putnam's Sons, 1962.

Lyall, Charles J. *Ancient Arabian Poetry.* London: Williams & Norgate, 1930.

Manguel, Alberto, and Gianni Guadalupi. *The Dictionary of Imaginary Places.* San Diego: Harcourt Brace Jovanovich, 1987.

Margoliouth, D. S. *The Relations between Arabs and Israelites prior to the Rise of Islam.* London: British Academy, 1921.

Marshall, Brian. "Bertram Thomas and the Crossing of al-Rub' al-Khali," in *Arabian Studies* 7 (1985).

Master of Belhaven (A. Hamilton). *The Kingdom of Melchior.* London: John Murray, 1949.

Matheson, Sylvia A. *Persia: An Archaeological Guide.* Park Ridge, N.J.: Noyes Press, 1973.

Maududi, S. Abdul a'la. *The Meaning of the Quran.* Lahore, Pakistan: Islamic Publications, 1986.

McCauley, John F., et al. "Subsurface Valleys and Geoarchaeology of the Eastern Sahara Revealed by Shuttle Radar," *Science* 218 (Dec. 3, 1982).

McClure, Harold A. "Radiocarbon Chronology of Late Quaternary Lakes in the Arabian Desert," *Nature* 263, no. 5580 (Oct. 28, 1976).

Meulen, Daniel van der. *Aden to the Hadramaut.* London: John Murray, 1947.

Meulen, Daniel van der, and H. von Wissmann. *Hadramaut: Some of Its Mysteries Unveiled.* Leyden: E. J. Brill, 1932.

Miller, Konrad. *Arabischewelt und Landerkarten.* Stuttgart: Selbstverlag des Herausgebers, 1927.

Minorsky, V., trans. *Hudud al-Alam, "The Regions of the World."* Karachi: Indus Publications, 1980.

Miquel, André. *La géographie humaine du monde musulman jusq'au milieu du 11c siècle.* Paris: La Haye, Mouton, 1967.

Morris, James. *Sultan in Oman.* London: Century Travellers, 1986.

Moscati, Sabatino. *Ancient Semitic Civilization.* New York: G. P. Putnam's Sons, 1960.

Moutsopoulos, N. C. "Observations sur les représentations du panthéon nabateen," in Fawzi Zayadine, ed., *Petra and the Caravan Cities.* Amman, Jordan: Department of Antiquities, 1990.

Müller, Walter W. "Notes on the Use of Frankincense in South Arabia," in *Proceedings of the Ninth Seminar for Arabian Studies.* London: Seminar for Arabian Studies, 1976.

———. "Arabian Frankincense in Antiquity According to Classical

Sources," in *Studies in the History of Arabia*, vol. 1. Riyadh, Saudi Arabia: University of Riyadh, 1977.

———. "Names of Aromata in Ancient South Arabia." Paper delivered at Convergo I Profumi d'Arabia, Pisa, Italy, 1995.

Naval Intelligence Division. *Western Arabia and the Red Sea*. London: Naval Intelligence Division, 1946.

Newby, Gordon D. *A History of the Jews of Arabia*. Columbia, S.C.: University of South Carolina Press, 1988.

Nicholson, Reynold A. *A Literary History of the Arabs*. Cambridge, Eng.: Cambridge University Press, 1966.

Nielsen, D., et al. *Handbuch de altarabischen Altertumskunde*. Copenhagen: Nyt Nordisk Forlag, 1927.

Noja, Sergio, ed. *L'Arabie avant L'Islam*. Aix-en-Provence, France: Edisud, 1994.

Norris, H. T. Entry on Islam, in *An Illustrated Encyclopedia of Mythology*. New York: Crescent Books, 1980.

Oldfather, C. H., trans. *Diodorus of Sicily*. Cambridge, Mass.: Harvard University Press, 1979.

O'Leary, De Lacy. *Arabia Before Muhammad*. London: Kegan Paul, Trench, Trubner, 1927.

O'Shea, Raymond. *The Sand Kings of Oman*. London: Methuen, 1947.

Overstreet, William C., Maurice Grolier, and M. Toplyn, eds. *The Wadi Al-Jubah Project: Geological and Archaeological Reconnaissance in the Yemen Arab Republic*. Washington: American Foundation for the Study of Man, 1988.

Palgrave, William Gifford. *Narrative of a Year's Journey through Central and Eastern Arabia*. London: Macmillan, 1865.

Philby, Harry St. John. *The Heart of Arabia*. London: Constable, 1922.

———. *The Empty Quarter*. New York: Henry Holt, 1933.

———. *Arabian Highlands*. Ithaca, N.Y.: Cornell University Press, 1952.

Phillips, Wendell. *Qataban and Sheba*. New York: Harcourt, Brace, 1955.

———. *Unknown Oman*. London: Longmans, 1966.

Pirenne, Jacqueline. "The Incense Port of Moscha (Khor Rori) in Dhofar," *Journal of Oman Studies* 1 (1975; Ministry of Information and Culture, Sultanate of Oman).

———. "Das Grecs à l'aurore de la culture monumentale Sabéene," in T. Fahd, ed., *L'arabe preislamique et son environment historique et cultural*. Strasbourg: Université des Sciences Humaines, 1989.

Potts, Daniel T. "Transarabian Routes of the Pre-Islamic Period," in J-F. Salled, ed., *L'Arabie et ses Mers Bordières*. Paris: GS — Maison de l'Orient, 1988.

Price, David. *Essay toward the History of Arabia*. "Arranged from the Tarikh Tebry and other authentic sources." London: printed by the author, 1824.

Rackham, H., trans. *Pliny: Natural History*. Cambridge, Mass.: Harvard University Press, 1986.

Rapoport, Amos. *Environment and Culture*. New York: Plenum Press, 1980.

Rappoport, Angelo S. *Ancient Israel: Myths and Legends*. New York: Bonanza Books, 1987.

Reich, Rosalie, ed. *Tales of Alexander the Macedonian*. New York: Ktav, 1972.

Rice, David T. *The Illustrations to the "World History" of Rashid al-Din*. Edinburgh: University of Edinburgh Press, 1980.

Rice, Edward. *Captain Sir Richard Francis Burton*. New York: Charles Scribner's Sons, 1990.

Ripinsky, Michael M. "The Camel in Ancient Arabia," *Antiquity* 49, no. 196 (Dec. 1975).

Robin, Christian. "The Rise and Fall of Ancient Kingdoms," in C. Desjuenes, ed., *Version Originale. Le Trimestriel de Réflex-*

ion. *The Arabian Peninsula*, no. 3. Paris: Version Originale, 1993.

Rosen, Arlene M. *Cities of Clay: The Geoarchaeology of Tells*. Chicago: University of Chicago Press, 1986.

Ryckmans, Jacques. "De l'or, de l'encens et de la myrrhe," *Revue biblique* 58 (1951).

————. "An Ancient Stone Structure for the Capture of Ibex in Western Saudi Arabia," in *Proceedings of the Ninth Seminar for Arabian Studies*. London: Seminar for Arabian Studies, 1976.

————. "Religion of South Arabia," in David N. Freedman, ed., *The Anchor Bible Dictionary*, vol. 6. New York: Doubleday, 1992.

————. "Inscribed Old South Arabian Sticks and Palm-leaf Stalks: An Introduction and a Palaeographical Approach," in *Proceedings of the Twenty-Sixth Seminar for Arabian Studies*. London: Seminar for Arabian Studies, 1993.

Sale, George, ed. *The Koran*. London: Thomas Tegg and Son, 1838.

al-Saleh, Khairat. *Fabled Cities, Princes and Jinns* (London: Peter Stone, 1985.

al-Sayari, Saad S., and Josef G. Zotl. *Quaternary Period in Saudi Arabia*. Vienna: Springer-Verlag, 1984.

Sedov, Alexander V., and Ahmad Batayi. "Temples of Ancient Hadramawt," in *Proceedings of the Twenty-Seventh Seminar for Arabian Studies*. London: Seminar for Arabian Studies, 1994.

Selincourt, Aubrey de, trans. *Herodotus: The Histories*. New York: Penguin Books, 1985.

Serjeant, Robert B. "Hud and Other Pre-Islamic Prophets of Hadramawt," *Le Museon* 57 (1954).

————. *South Arabian Hunt*. London: Luzac, 1993.

————. "Zina: Some Forms of Marriage and Allied Topics in Western Arabia," in Andre Gingrich et al., eds., *Studies in Oriental Culture and History*. Frankfurt: Peter Lang, 1993.

Serjeant, Robert B., and Ronald Lewcock. *San'a: An Arabian Islamic City*. London: World of Islam Festival Trust, 1983.

Shad, Abdul Rehman. *From Adam to Muhammad*. Lahore, Pakistan: Kazi, 1987.

al-Shahari, Ali Achmed Mahash. "Grave Types and 'Trilith' in Dhofar," *Arabian Archaeology and Epigraphy* 2 (1991).

Shahid, Irfan. *Byzantium and the Arabs in the Fourth Century*. Washington: Dumbarton Oaks Research Library and Collections, 1984.

Smith, G. Rex. "Ibn al-Mujawir on Dhofar and Socotra," in *Proceedings of the Eighteenth Seminar for Arabian Studies*. London: Seminar for Arabian Studies, 1985.

———. "Some 'Anthropological' Passages from Ibn al-Mujawir's Guide to Arabia and Their Proposed Interpretations," in Andre Gingrich et al., eds., *Studies in Oriental Culture and History*. Frankfurt: Peter Lang, 1993.

Smith, Watson. "One Man's Archaeology," *Kiva* 57, no. 2 (1992).

Smith, William. *An Atlas of Ancient Geography*. London: John Murray, 1874.

Sprenger, A. *Die Alte Geographie Arabiens*. Amsterdam: Meridian, 1966.

Stark, Freya. *Seen in the Hadhramaut*. New York: E. P. Dutton, 1939.

———. *The Southern Gates of Arabia*. Los Angeles: J. P. Tarcher, 1983.

Stevenson, Edward L., trans. *Claudius Ptolemy: The Geography*. New York: Dover, 1991.

al-Taee, Nasser. "Oman: An Architecture," in *Pride*. Muscat, Oman: Alroyz Publishing, 1994.

Tey, Josephine. *The Singing Sands*. New York: Macmillan, 1952; reprint, Collier Books, 1988.

Thackston, W. M., Jr., trans. *The Tales of the Prophets of al-Kisa'i*. Boston: Twayne, 1978.

Thesiger, Wilfred. "A New Journey in Southern Arabia," *Geographical Journal* 10, nos. 4–6 (Oct.–Dec. 1946).

———. "Across the Empty Quarter," *Geographical Journal* 111, nos. 1–3 (Jan.–Mar. 1948).

———. *Arabian Sands.* New York: E. P. Dutton, 1959.

Thomas, Bertram. *Alarms and Excursions in Arabia.* Indianapolis: Bobbs-Merrill, 1931.

———. "A Journey into Rub' al Khali — the Southern Arabian Desert," *Geographical Journal* 77, no. 1 (Jan. 1931).

———. "A Camel Journey across the Rub' al Khali," *Geographical Journal* 78, no. 3 (Sept. 1931).

———. *Arabia Felix: Across the "Empty Quarter" of Arabia.* New York: Charles Scribner's Sons, 1932.

———. *The Arabs.* Garden City, N.Y.: Doubleday, Doran, 1937.

———. "Four Strange Tongues from Central South Arabia — the Hadara Group," in *Proceedings of the British Academy, 1937.* London: British Academy, 1938.

Thurman, Sybil. "Rivers of Sand," *Aramco World Magazine* 34, no. 6 (Nov.–Dec. 1983).

Tibbetts, G. R. *Arabia in Early Maps.* Cambridge, Eng.: Falcon-Oleander, 1978.

Tidrick, Kathryn. *Heart-Beguiling Arabia.* Cambridge, Eng.: Cambridge University Press, 1981.

Tkach, J. Entry for Iobartai, in G. Wissowa, ed., *Paulys Real-Enzyclopädie der Klassischen Altertumwissenschaft.* Stuttgart: J. B. Metzler, 1916.

Trench, Richard. *Arabian Travellers.* Topsfield, Mass.: Salem House, 1986.

Trimingham, J. Spencer. *Christianity among the Arabs in Pre-Islamic Times.* London: Longman, 1979.

Turnquist, Gary M. "The Pillars of Hercules Revisited," *Bulletin of the American Schools of Oriental Research*, no. 216 (Dec. 1974).

Van Beek, Gus W. "Frankincense and Myrrh," *Biblical Archaeologist* 23, no. 3 (Sept. 1960).

——. *Hajar bin Humeid.* Baltimore: Johns Hopkins University Press, 1969.

——. "The Rise and Fall of Arabia Felix," *Scientific American* 221, no. 6 (Dec. 1969).

——. "South Arabian History and Archaeology," in *The Bible and the Ancient Near East: Essays in Honor of William Foxwell Albright.* Winona Lake, Ind.: Eisenbrauns, 1979.

Vycichl, Werner. "Studies in Nabataean Archaeology and Religion," in Fawzi Zayadine, ed., *Petra and the Caravan Cities.* Amman, Jordan: Department of Antiquities, 1990.

Ward, Philip. *Travels in Oman: On the Track of Early Explorers.* Cambridge, Eng.: Oleander Press, 1978.

Whalen, Norman M., and David W. Pease. "Early Mankind in Arabia," *Aramco World* 43, no. 4 (July–Aug. 1992).

Wherry, E. M. *A Comprehensive Commentary on the Quran.* London: Kegan Paul, Trench, Trubner, 1896.

Williams, Robin J. "In Search of a Legend — the Lost City of Ubar," *P.O.B.* 17, no. 6 (Aug.–Sept. 1992).

Wolohojian, Albert M., trans. *The Romance of Alexander the Great by Pseudo-Callisthenes.* New York: Columbia University Press, 1969.

Wrede, Adolph von. *Reise in Hadhramaut.* Braunschweig: Verlag von Friedrich Bieweg und Sohn, 1870.

Wright, John K. *The Geographical Lore at the Time of the Crusades.* New York: American Geographical Society, 1925.

Wustenfeld, Ferdinand. *Jacut's Geographisches Wörterbuch.* Leipzig: Bei F. A. Brockhaus, 1869.

Yar-Shater, Edshan, ed. *The History of al-Tabari.* Albany: State University of New York Press, 1989.

Zarins, Juris. "The Camel in Ancient Arabia: A Further Note," *Antiquity* 52 (1978).

————. "The Transarabia Expedition: Archaeological and Historical Perspectives." Manuscript, 1990.

————. "Pastoral Nomadism in Arabia: Ethnoarchaeology and the Archaeological Record — a Case Study," in O. Bar-Yosef and A. Kahznov., eds., *Pastoralism in the Levant*. Madison, Wis.: Prehistory Press, 1992.

————. "Prehistory of Arabia," in David N. Freedman, ed., *The Anchor Bible Dictionary*, vol. 1. New York: Doubleday, 1992.

————. "Dhofar — Land of Incense: Archaeological Work in the Sultanate of Oman, 1990–1995." Manuscript, 1996.

Zarins, Juris, M. Ibrahim, D. Potts, and C. Edens. "Saudi Arabian Archaeological Reconnaissance 1978: The Preliminary Report on the Third Phase of the Comprehensive Archaeological Survey Program — the Central Province," *Atlal* 3 (1979).

Zarins, Juris, N. Whalen, M. Ibrahim, A. Mirsi, and M. Khan. "Preliminary Report on the Central and Southwestern Provinces Survey: 1979," *Atlal* 4 (1980).

Zarins, Juris, A. Murad, and K. Al-Yish. "The Second Preliminary Report on the Southwestern Province," *Atlal* 5 (1981).

Zarins, Juris, A. S. Mughannum, and M. Kamal. "Excavations at Dhahran South — the Tumuli Field," *Atlal* 8 (1984).

Zohar, M. "Pastoralism and the Spread of the Semitic Languages," in O. Bar-Yosef and A. Kazanov, eds., *Pastoralism in the Levant*. Madison, Wis.: Prehistory Press, 1992.

Acknowledgments

Along with the individuals mentioned in the text, a legion of friends made the Ubar adventure possible.

For inspiring an early interest in archaeology, I'm grateful to archaeologist and family friend Watson Smith. He took the time to tell me, a little kid, what he had discovered in the lands of the Zuni and Hopi and how he hauled a piano to his site of Awatovi so his diggers could dance in the desert moonlight.

For forever tripping over maps and books spread out on the living room floor, I offer heartfelt thanks to my wife and expedition manager, Kay, and my daughters, Cristina and Jennifer. They and my mother, Helen Clapp, were unwavering in their encouragement.

In the project's research phase, I benefited from the advice of UCLA archaeologists Giorgio and Marilyn Buccelati, and Dunning Wilson of the University Research Library. I was welcomed at the Huntington Library by the curator of rare books, Alan Jutzi, and the director, Bill Moffit. And I'm grateful to the good people at the Oriental Collection of the New York Public Library, the British Library, the University of Edinburgh Library, and Cambridge University's Institute of Oriental Studies.

When it came to translating arcane texts, Bill Moritz and Klaus and Gabi Brill guided me through the swamps of nineteenth-century academic German; Paul Boorstin resurrected his college French; Nasser al-Taee helped me with modern Arabic; and UCLA's Mohammed Shikrila and Caltech's Anthony Aebi translated medieval Arabic.

When it looked as though an expedition might actually be possible, Judy Miller was enormously helpful, as were Len Berlin and Bill and Beth Overstreet, U.S. Geological Survey veterans of Arabia. A special thanks is due Gordon and Merilyn Hodgeson of the American Foundation for the Study of Man. And gold stars for moral support go to Harold and Judy Hayes and John and Louise Brinsley.

In the Sultanate of Oman, our initial sponsors were the Oman International Bank, presided over by H.E. Dr. Omar Zawawi and John Wright; the al-Bustan Palace Hotel, managed by Chris Cowdray; Gulf Air; and Occidental Petroleum, whose Gene Grogan, Derek Hart, and Dr. Armand Hammer all took great personal interest in the project.

Additional expedition sponsors were Airwork, Limited (and John Fulford); American Airlines; Avis Car Rental; British Petroleum Oman; Desert Line / Sheikh Ahmed Farid; DHL Oman; Oman's General Telecom Organization; Genetco; IBM / Gulf Business Machines; Land Rover / Mohsid Haider Darwish; Matrah Cold Stores; Oman's Ministry of Information and Ministry of National Heritage and Culture; Nortech; Oman Aviation Services; Oman National Insurance; Oman United Agencies; Petroleum Development Oman; Rowntree Mackintosh; Royal Insurance of Oman; Royal Oman Police; Salalah Holiday Inn; Smith Kline Beecham; Suhail and Saud Bahwan; Thomson CSF; United Media Services; and W. J. Towell (and Kamal Sultan).

After the expedition, filmmaking colleagues Rob Bogdanoff, Jennifer Dolce, Harrison Engle, Irwin Rosten, David Saxon, Terry Sanders, and Mel Stuart gave of their time and valued advice. Further archaeological guidance was provided by Artemis and Martha Joukowsky, of the American Center of Oriental Research and Brown University's Center for Old World Art and Archaeology, respectively.

And friends Ed and Cynthia Lasker became so intrigued by Ubar that they traveled to Oman to see the once-wicked city for themselves.

Years ago, good friends Bob Ivey and Jill Bowman were the first to suggest that if anything ever came of the Ubar project, I should write a book. A lawyer, Bob acted as my agent and recommended the book to Houghton Mifflin's Harry Foster, who, working with Peg Anderson, edited this manuscript and saw it through to publication. I thank them for their great care and deft touch. And I thank Kristen Mellen and Anne Chalmers for their admirable realization of the book's graphics.

Index

NOTE: page numbers in italics refer to illustrations.